G. Sherouse

PRACTICAL ASPECTS OF ELECTRON BEAM TREATMENT PLANNING

EDITED BY
COLIN G. ORTON, Ph.D.
Rhode Island Hospital &
Brown University
and
FARIDEH BAGNE, Ph.D.
Duke University Medical Center
Durham, North Carolina

Proceedings of the "Practical Aspects of Electron Beam Treatment Planning" Symposium on July 31, 1977 in Cincinnati, Ohio, sponsored by the American Association of Physicists in Medicine Radiation Therapy/Computer Applications Joint Task Group on Electron Beams: Farideh Bagne, Ph.D., Chairman, Geoffrey S. Ibbott, Douglas Jones, B.S., James A. Purdy, Ph.D., and Larry D. Simpson, Ph.D.

Library of Congress Catalog No. 78–65599
International Standard Book No. 0-88318-247-5
International Standard Serial No. 0163-1802

Copyright 1978 by the American Association of
Physicists in Medicine

Published by the American Institute of Physics
335 East 45 Street, New York, New York 10017

Printed in the United States of America

Table of contents

THE AAPM MONOGRAPH SERIES J.S. Laughlin	1
INTRODUCTION F. Bagne	3
SYMPOSIUM WELCOMING ADDRESS P. Wootton	7
1. DOSIMETRY OF ELECTRON BEAMS N. Suntharalingam	11
1.1. Introduction	11
1.2. Dosimetric parameters	11
1.3. Energy	12
1.4. Skin dose	13
1.5. Depth dose	14
1.6. Beam uniformity	17
1.7. Isodose distributions	19
1.8. Summary	20
1.9. References	20
2. PATIENT-SPECIFIC DATA ACQUISITION IN ELECTRON BEAM TREATMENT PLANNING J. Purdy	22
2.1. Introduction	22
2.2. Contour	22
2.3. Normal structure localization	24
2.4. Tumor localization	26
2.5. Inhomogeneities	27
2.6. Field shaping	27
2.7. Simulation	30
2.8. References	31
3. TREATMENT PLANNING CONSIDERATIONS J. Ovadia	33
3.1. Introduction	33
3.2. Central axis depth doses	33
3.3. Single field isodose distributions	36

3.4. Adjacent fields	37
3.5. Electron treatment planning program	41
3.6. Combination of fields and tumor dose specifications	42
3.7. Tissue heterogeneity	44
3.8. Contaminant neutron dose	48
3.9. Integral dose	49
3.10. References	50

4. ALGORITHMS FOR COMPUTERIZED TREATMENT PLANNING — 52
E. Sternick

4.1. Introduction	52
4.2. Empirical approaches	53
4.2.1. Absorption equivalent thickness (AET) method	53
4.2.2. Absorption coefficient method	54
4.2.3. CET method	54
4.2.4. MAC method	56
4.2.5. Equivalent thickness method and edge effects	58
4.2.6. Calculations for pendulum therapy	60
4.3. Semi-empirical approaches	62
4.3.1. Age diffusion method	62
4.3.2. Pencil beam approximation	63
4.3.3. The formula of Czaikawsky	63
4.3.4. The difference method	64
4.4. Analytical approaches	65
4.4.1. Central axis calculation	65
4.4.2. Pencil beam method	65
4.5. Conclusions	67
4.6. References	67

5. MEMORIAL ELECTRON BEAM AET TREATMENT PLANNING SYSTEM — 70
J.G. Holt *et al.*

5.1. Introduction	70

5.2. Computation model	71
5.3. Measurements	74
5.4. References	78
6. LOW-ENERGY ELECTRONS F. Bagne and M. Tulloh	80
6.1. Introduction	80
6.2. Dosimetric considerations	80
6.3. Treatment of skin tumors	83
6.4. Tumors near skin surface	84
6.5. Total body irradiation	84
6.6. Treatment of chest wall	89
6.7. Bolus material	93
6.8. References	94
7. DISCUSSION	97
8. INDEX	107

The AAPM Monograph Series

With this monograph "Practical Aspects of Electron Beam Treatment Planning," the Publications Committee is carrying out its established policy of publishing a series of monographs for the American Association of Physicists in Medicine based on the proceedings of our AAPM Summer Schools, Symposia during our meetings, and workshops carried out under AAPM sponsorship. The subjects of these various meetings are timely, and competent individuals have contributed substantially in thought and time to their presentations in them. Accordingly, in order to accommodate expressed scientific and professional interest, this series of edited monographs of the proceedings of these activities of our Association is being made available.

The first in this series was on Technetium-99m edited by James Kereiakes. *This* monograph is on the subject of electron beam dosimetry and is edited by Colin Orton and Farideh Bagne. High energy electron beams have become an established treatment modality since their development over a quarter of a century ago, and our Publications Committee is pleased to be able to provide this monograph on this most important subject.

John S. Laughlin
September 2, 1978

Introduction

This monograph is a compilation of papers presented at a symposium held as part of the Annual Meeting of the American Association of Physicists in Medicine on July 31, 1977 in Cincinnati, Ohio. The objectives of the conference were: 1) to review the current state of electron beam dosimetry and treatment planning, 2) to provide practical information and guidelines for clinical Medical Physicists, and 3) to indicate directions for future development in this area.

Although advantageous in specific clinical cases, electrons pose certain difficulties in dosimetry and treatment planning. The body inhomogeneities such as lung, bone, and oral cavities, as well as any additional compensators, field shaping blocks, or wedges in the path of the beam can create remarkable variations in dose distribution within the treatment volume. If such dose non-uniformity is not realized, possible tumor recurrence or complications may follow. In order to obtain a uniform dose throughout the treatment volume without overdosing adjacent tissues, a thorough knowledge is needed of the particular properties of electrons and their absorption in various tissues. Also required is familiarity with problems in the dosimetry and treatment planning of electron beams.

For several years, the AAPM committees on Radiation Therapy and Computer Applications felt the need for establishing working groups dealing with the use of electron beams in radiation therapy. Edward S. Sternick, the former Chairman of the Computer Applications Committee Task Group No. 1 on "Guidelines on Computerized Electron Beam Treatment Planning," had suggested in 1975 that a symposium on electron beam treatment planning be held. Subsequently, a joint task group of the Radiation Therapy and Computer Applications Committees was established in 1976 to study the treatment planning aspects of electron beams and this symposium results.

It is with gratitude that the Task Group acknowledges the support and encouragement given to them by Kenneth Wright, Chairman of the Radiation Therapy Committee, and Colin Orton, Chairman of the Computer Applications Committee. The Task Group would also like to acknowledge the following members of the AAPM for their interest and comments: William R. Hendee, President; Edward W. Webster, Chairman, Publications Committee; Earle C. Gregg, Chairman, Science Council; Bengt E. Bjarngard, Chairman, Program Committee; N. Suntharalingam, Chairman, 1977 AAPM Meeting Scientific Program

Committee; James G. Kereiakes, Chairman of the Local Arrangements Committee, and Sharon Pierce, AAPM Executive Director.

Farideh Bagne, Ph.D.
Chairman
Task Group on Electron Beams
Radiation Therapy Committee,
American Association of Physicists in Medicine.

Symposium: Welcoming address

Ladies and Gentleman, on behalf of the Directors and Officers of the AAPM, I have the honor and pleasure of welcoming you to this symposium on "The Practical Aspects of Electron Beam Treatment Planning".

The use of fast electrons in radiation therapy may be considered to have originated with radium beta ray plaques. However, they were capable of adequately treating only the first few millimeters of tissue.

Machine generated kinetic electrons emerging from a Lenard-type tube after acceleration in a field of a few hundred thousand volts were used for dermatological conditions in Germany during the 1930s.

Significant beam intensities and penetration for serious evaluation of the modality only became available with the development of methods of accelerating electrons to high energies either by rotational machines such as betatrons and synchrotrons or by linear devices such as Van de Graaff and UHF powered linear accelerators. The protagonists of the rotational machines solved the technical problems of beam extraction and all faced the problem of producing an effectively wide beam. These developments occurred and the technical problems were solved during the decade 1943–1953 principally in the United States, Germany and Switzerland. The development of these machines provided a major impetus to the profession of Medical Physics in this country. A list of the authors of papers on the physical aspects of electron beam therapy reads like an honor roll of the AAPM. The following decade saw the development of the dosimetry and other techniques appropriate to the application of the modality to practical therapy culminating in the Symposium on High Energy Electrons, Montreux, Switzerland, 1964.

Since that date, there have been several gatherings at which electrons have been linked with high energy photons or as alternative or supplementary therapy modalities, such as the "High Energy Radiation Therapy Dosimetry" Conference held by the National Academy of Sciences in 1967, and the International Symposium, "The Clinical Usefulness of High Energy Photons" at Thomas Jefferson University in 1975.

Formal codes of calibration practices were drawn up in 1966 and others in 1971 and 1972. Refresher courses on the use of electron beams have been a feature of the RSNA for some years.

Why the need for another electron beam symposium? Despite these activities, for many of us electron beam therapy has been a spectator sport. The recent surge in sale and installation of high energy linear accelerators having photon *and* multi-energy electron beam capabilities has changed this circumstance. Electron beam therapy is no longer an esoteric modality confined to relatively few institutions so that it now behooves every member of AAPM with clinical pretensions to be familiar with the working details as well as the broad principles of its dosimetry. We need to be authoritative on planning methods and to understand the state of the treatment planning art. When we study the details of the calibration protocols, we find that there are differences that need to be reconciled. There are more physicists with a wider variety of instrumentation making more accurate measurements and thus encountering different and sometimes more subtle problems that need to be discussed.

The organizers of this symposium are to be congratulated in having recruited representatives of each phase of the development of electron beam therapy with others of that august group in the audience, and I, for one, look forward to an educational experience, and updating on the state-of-the-art, some resolution of differences and above all, a stimulating discussion.

Welcome to the Symposium on Practical Aspects of Electron Beam Treatment Planning.

> Peter Wootton, President-Elect
> American Association of Physicists
> in Medicine

1. Dosimetry of electron beams

Nagalingam Suntharalingam, Ph.D.

Thomas Jefferson University Hospital, Philadelphia, Pennsylvania

1.1. INTRODUCTION

Though in clinical use for almost twenty-five years,[1] electron beam treatment planning and dose computational methods have not been developed to the same degree of sophistication and practicality as for photon beams. This is due, at least in part, to the generally accepted concept that most electron beam treatments are delivered using single fields and are usually confined to superficial tumors. During the past few years there has been a dramatic change and electron beams are now considered indispensable for various clinical situations.[2] Isocentrically mounted small electron accelerators with both photon and electron modalities and clinically useful ranges of energies and dose rates are now a reality. Precise localization of tumor and critical organs is now possible due to the recent advances in tomographic scanning. The effectiveness of treatment planning and dose distribution computations has been increased as a result of the use of digital mini-computers. All these recent technological advances have enhanced the development of new treatment techniques using electron beams, and as such more careful and detailed dosimetric studies on electron beams are now essential.

In this presentation some basic dosimetric parameters necessary for electron beam treatment planning will be reviewed. Methods of measurement of these physical parameters, including a discussion of certain errors and limitations of the different techniques, and some representative quantitative data useful for characterization of electron beams will also be given.

1.2. DOSIMETRIC PARAMETERS

The most useful dosimetric parameters that adequately characterize the physical properties of an electron beam are:
- skin or surface dose
- initial build-up of dose with depth
- depth of dose maximum
- fall-off of dose with depth
- uniformity of dose across the beam
- central plane isodose distributions

1.3. ENERGY

All these parameters are dependent on the kinetic energy of the electrons, and hence it is important to clearly specify and define the single energy parameter chosen to characterize the electron beam. The most commonly used specification is the energy at the incident surface E_o, and it is very often determined using well established empirical relationships with the practical range R_p. For water, the Markus equation,[3]

$$R_p = 0.521 E_o - 0.376,$$

is valid within 2% for electron energies from about 5 to 40 MeV. The practical range in water of an electron beam can be determined using small volume ionization chambers, preferably flat detectors of the "pancake" type. The measured ionization values are plotted as a function of depth, and the practical range is defined as that depth where the extrapolation of the straight descending portion of the curve meets the extrapolation line of the Bremsstrahlung background (Fig. 1.1). The measured practical range depends on the measuring geometry, and a recommended method is to use a beam diameter greater than the practical range, and a detector with a very small diameter. Corrections for beam divergence, as the detector is moved towards greater depths, have also to be taken into consideration. Precise values of the practical range will be required for some of the algorithms used in the generation of electron beams and in correcting for inhomogeneities.

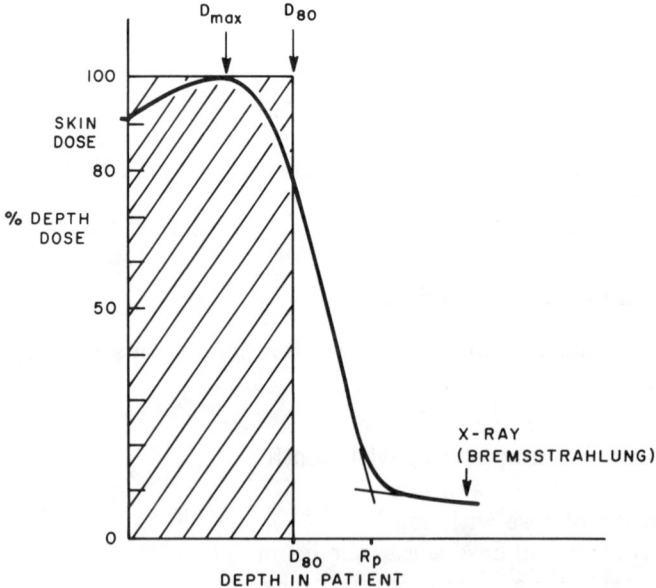

FIG. 1.1. Electron beam depth-dose distribution.

1.4. SKIN DOSE

The skin or surface dose from electron beams is typically in the range of 85–95% of the dose at the depth of maximum build-up, and hence electron beams have little or no skin-sparing properties. The higher the electron beam energy the higher the skin dose. The presence of scattering foils, transmission chambers, collimators, cones, etc. introduce low energy scattered electrons into the incident beam and have the effect of increasing the surface dose and moving the maximum towards the surface.[4] Electron beams generated using scanning magnets and open collimators give less skin doses because the scanning mechanism does not affect the beam energy or introduce low energy electrons into the beam.[5]

Skin dose or surface doses are best measured using extrapolation or build-up ionization chambers. Thin thermoluminescent dosimeters, such as the LiF-Teflon discs, are also useful. Typical results of measured surface doses from electron beams from an A-45 betatron are given in Table 1.1. Shown for comparison are the surface doses from different photon beams.

TABLE 1.1. Some typical surface doses for high energy beams using LiF-Teflon disk dosimeters.

	Field Size	Collimator-Surface Distance (cm)	Surface Dose (% max)
		PHOTONS	
Cobalt-60	10×10	32.5	20
(AECL-Theratron 80)	33×33	32.5	60
4 MV x rays	10×10	40	18
(Varian Clinac-4)	32×32	40	50
45 MV x rays	5×5	28	12
(Brown Boveri	10×10	28	20
Betatron)	20×20	28	24
		ELECTRONS	
5 MeV	20×20 scanning	no electron cone	70
5 MeV	14×10	electron cone	85
45 MeV	14×10	electron cone	95

1.5. DEPTH DOSE

The absorbed dose distribution with depth along the central axis of an electron beam is represented by the curve shown in Fig. 1.1. Any attempt at characterizing an electron beam should adequately allow for the important features of this distribution, namely the initial build-up region, the depth of dose maximum, the fall-off of dose with depth, and the tail of the curve resulting from the photon contamination. The one parameter that most influences the shape of this dose distribution is the electron beam energy. The presence of degrading materials such as scattering foils, transmission chambers, collimators, and air in the path of the incident beam, influences the depth-dose curve by decreasing and broadening the energy distribution, by producing low energy electrons and large angle scattered electrons and by also producing Bremsstrahlung photons. The introduction of the lower energy scattered electrons into the beam has the effect of displacing the depth of dose maximum towards the surface. The shape and gradient of the fall-off portion of the depth-dose curve is also affected by collimation and scattering of the electron beam.

The magnitude of the photon contamination depends on the material and thickness of the scattering foil used. When the photon back-

TABLE 1.2. Depth of 80% dose levels for different electron beams.

	Energy (MeV)	80% Depth (cm)
SIEMENS BETATRON (50 cm TSD)		
	7	1.8
	9	2.6
	11	3.2
	15	4.4
	18	4.7
SAGITTAIRE-LINAC (100 cm TSD)		
	7	2.1
	19	6.7
BBC-35 MeV BETATRON (100 cm TSD)		
	25	7.3
	35	10.7
BBC-45 MeV BETATRON (110 cm TSD)		
	5	1.6
	10	3.2
	25	7.7
	35	8.8
	45	9.9

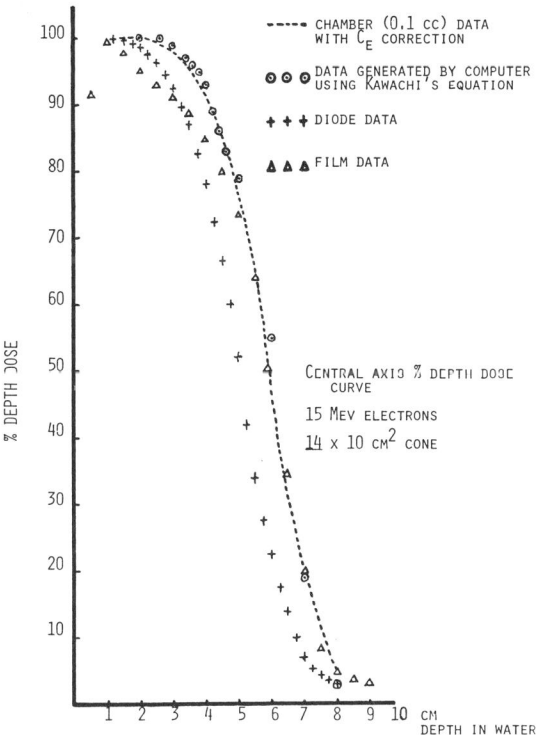

FIG. 1.2. Comparison of % depth-dose data generated using film, 0.1 cc ion-chamber, diode probe, and analytical (modified Kawachi's equation) methods.

ground is in the region of 5 to 10% it affects the depth of dose maximum and the gradient of the fall-off portion of the curve.

These difficulties or negative aspects of the use of scattering foils have been overcome by the use of quadrupole scanning magnets to scan, in a random fashion, the initial electron beam. The resultant beam is uniform, the beam energy is not affected, and no low energy electrons are produced.

The characteristics of electron beams which are of particular importance and provide the basis for their use in radiation therapy are the rapid build-up of dose and the sharp fall-off of dose with depth. The higher the energy the more rapid is the superficial build-up and the beam shows a broad maximum. The depth of dose maximum, usually in the range of 1–3 cm in water, corresponds to differences in mean energies and shows differences from one machine to another. The gradient of the sharp fall-off region decreases as the energy increases. Of special interest for clinical applications is the depth of the 80% dose level, and measured data for different machines and different beam energies are given in Table 1.2.

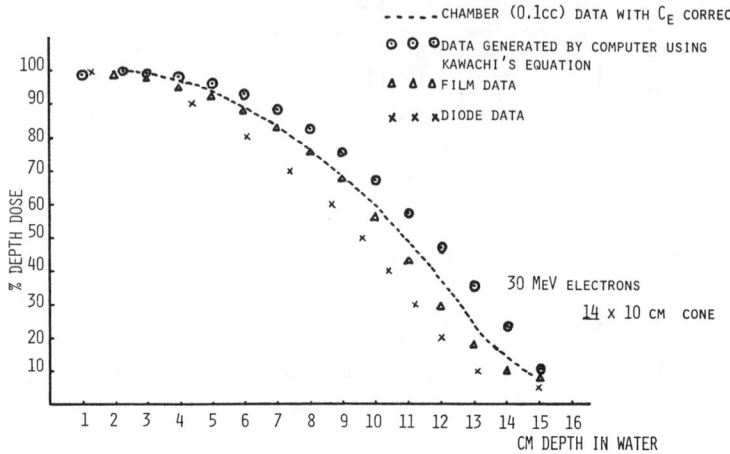

FIG. 1.3. Comparison of % depth-dose data generated using film 0.1 cc ion-chamber, diode probe, and anlytical (modified Kawachi's equation) method.

Central axis depth-dose curves can be measured using a variety of radiation dosimeters such as small volume ionization chambers, films, diodes, chemical dosimeters and TLD. Since the depth-dose curves are relative measurements, usually normalized to the dose value at maximum, special attention should be given to the absolute calibration of the beam. The need to develop uniform methods of absolute calibration of electron beams has led to a study of using photon calibrated ionization chambers using the so-called C_E method.[6] There still remains some uncertainty to this method and the correctness of the recommended C_E values. This problem is being addressed once more at both the national and international level, and one should be hearing more about this in the coming years.

Since electron beams behave differently from photon beams, in that the beam energy changes with depth, the energy response of the different dosimeters should be well established and the appropriate correction factors applied so as to arrive at the same data irrespective of the method used. As an example of the observed differences between different dose measuring systems, the relative responses of an ion chamber, film, and diode used to measure the depth dose of 15 MeV and 30 MeV electron beams respectively are shown in Figs. 1.2 and 1.3.

The recommended C_E dose conversion factors for the ion chamber responses have been applied. Shown for comparison is the data generated using an analytical expression (modified Kawachi's equation[7]). Dose conversion factors for the other systems are not available. It should be noted that the response and behavior of all diodes and all films are not identical. Care should be exercised in interpreting the results obtained using any one system. Statements indicating that solid-state diode detectors and film require no corrections to obtain relative depth dose curves are questionable.

The high spatial resolution obtainable with film makes this method very attractive and useful. Films have been and continue to be used to measure electron beams.[8] However, they do have problems attributable to the increased scatter associated with the high atomic number of the emulsion and the data for energies above about 20 MeV, especially in the build-up region, have been found to be unreliable. Consistent and reproducible results can be obtained if the dose-density response curve is determined separately for each experiment and for each batch of film. Care should also be given to the processing procedures.

Central axis depth-dose curves can also be measured using LiF-TLD.[9] As in the case of the use of film, this method has the practical advantage that only one exposure need be made. However, careful calibration procedures of the individual dosimeters will have to be followed to obtain reliable results.

1.6. BEAM UNIFORMITY

A uniform dose distribution across the treatment field is a desirable characteristic for clinical use of the electron beam. Field flatness or beam uniformity of electron beams have been achieved by the use of

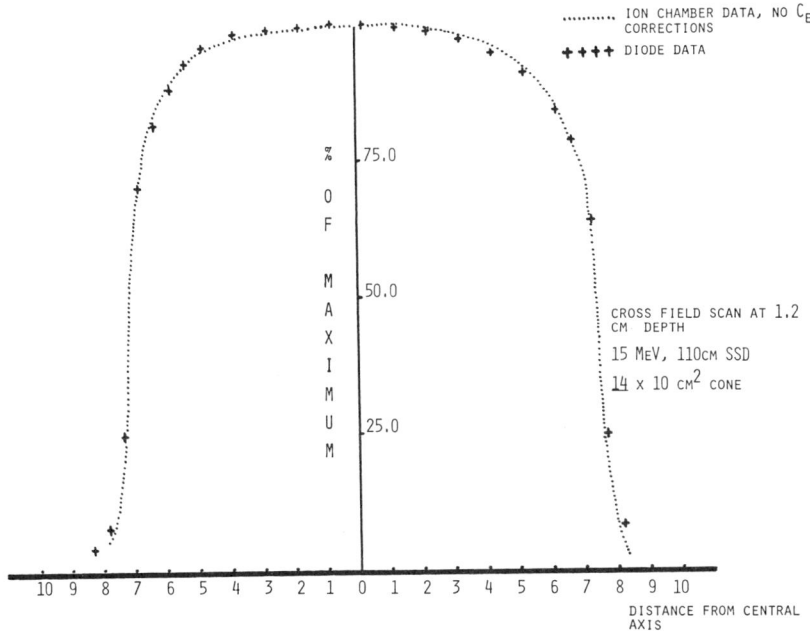

FIG. 1.4. Comparison of cross field scan data obtained using 0.1 cc ion-chamber and diode in water phantom.

scattering foils or quadrupole scanning magnets. Electrons scattered by the collimators also contribute substantially to achieving beam uniformity, more so at the shallower depths. In some machines the inner surface of the collimator has been aligned in such a way as to use the scattered electrons to raise the dose at the periphery of the field and improve beam uniformity. A ring of high density materials at the aperture of the collimator has also helped achieve better beam uniformity.

A method of assessing beam uniformity is to measure dose profiles across the beam at the different depths of interest. For clinical use, one usually accepts a beam to be uniform if the variation in dose at any point within 80% of the geometric beam area does not exceed by $\pm 5\%$. The Nordic Association of Clinical Physicists have defined a uniformity index[10] for the dose distribution of electron beams in a reference plane perpendicular to the beam axis. This index gives the ratio of the area contained by the 90% curve relative to the dose at the central ray and the geometric beam area. Good and acceptable beam uniformity is when the value of this index is greater than 0.8.

Figure 1.4 is an example of measured beam profiles using an ion chamber and diode for a 15 MeV electron beam. The data points are normalized to their individual response at the central axis. While the agreement for the relative values looks good, there exist uncertainties in the absolute dose values near the edges of the beam.

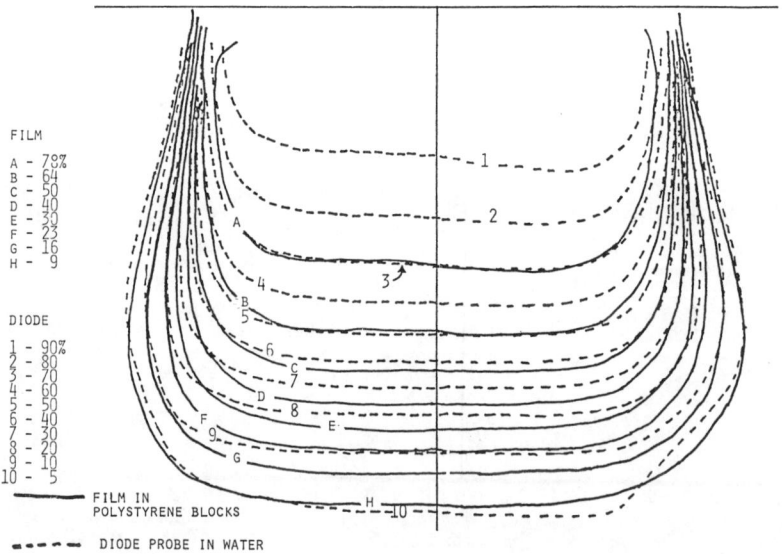

FIG. 1.5. Isodose curves of a 30 MeV electron field using film and diode methods.

1.7. ISODOSE DISTRIBUTIONS

The most commonly used dose distribution data for treatment planning purposes is the central plane isodose distribution. From this distribution one obtains information about the dose at off-axis points and more importantly the dose at the edges of the collimated beam. Treatment planning for electron beams, at least for the most part, has been straightforward since most treatments are delivered using single fields. Given the extent of the target volume to be irradiated, one usually chooses the appropriate energy and field size of the electron beam such that the 80% isodose curve surrounds the target volume.

There are several methods of obtaining electron beam isodose curves. The generation of isodose curves require a considerable number of data points, and as such one looks to automatic dose measuring and scanning devices. Diodes or ion chambers can be used with automatic isodose plotters to measure dose in water. Because of the variations in instantaneous dose-rate the relative response between two similar detectors should be utilized. Film methods do have a considerable advantage over all other systems because of its high spatial resolution and the relatively short machine time required to obtain the data.

Generally one makes a correction to adjust the central axis dose values to those obtained with a calibrated ion chamber, and accepts the off-axis points to also be in agreement. Figure 1.5 is an example of the central plane isodose curves of a 30 MeV electron beam measured using film and diodes. It can be seen that while the data along the central axis

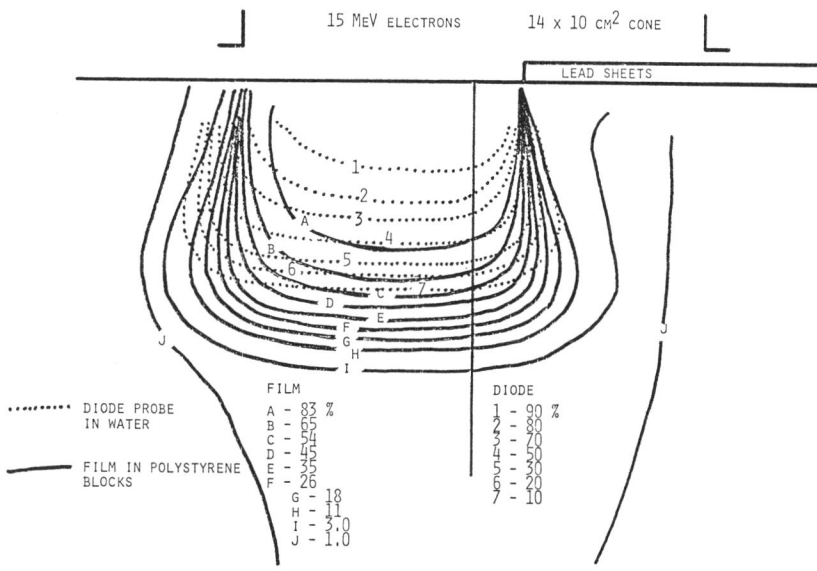

FIG. 1.6. Isodose curves of a 15 MeV blocked electron field using film and diode methods.

can be adjusted to agree to the ion chamber values, the differences at the beam edges are not insignificant. Using measured isodose curves, one gives important consideration to the "constriction" and "bulging" of the isodose curves at depth when choosing the appropriate field size for clinical use. These effects are due to the scattered electrons and the energy response of the detectors will have to be established and corrected for if a higher degree of accuracy is required. This may not be necessary for single field treatments but when one considers patching of two or more fields to irradiate large areas or, if one intends to generate summated distributions from multiple or moving beams, then the edge effects are important. The dramatic difference in the response of different detectors to scattered electrons can be seen in Fig. 1.6. Measurements with film compared to a diode indicate a very sharp dose fall-off at the edge of a field shaping lead block compared to the unblocked edge.

1.8. SUMMARY

In this presentation emphasis has been placed on those parameters that are important and necessary for electron beam dosimetry and treatment planning. The dose distribution for each single beam will have to be known accurately. The practical problems that arise in the characterization of the single electron beam have been addressed. There is a need for more input from physicists to better define and accurately measure the different parameters that represent the therapeutic beam. The measurements are necessary before one can obtain reliable treatment plans for techniques other than single field irradiations. Other factors, both external and patient related, also will influence the absorbed dose distributions under treatment conditions. All these considerations have to be carefully studied before one attempts a general method of computerized treatment planning for electron beam therapy.

1.9. REFERENCES

[1]H. Quastler, *el al.* "Techniques for the application of the betatron to medical therapy," Am. J. Roentg. 61, 591 (1949).
[2]N. duV. Tapley, *Clinical Applications of the Electron Beam* (Wiley, New York, 1976).
[3]B. Markus, "Energiebestimmung schneller elektronen aus tiefendosiskurven," Strahlentherapie 116, 280 (1961).
[4]H. Svensson and G. Hettinger, "The influence of the collimating system on the dose picture from 10 to 35 MeV electron radiation," Acta Radiologica 6, 404 (1967).
[5]P.R. Almond, "Dosimetry considerations of electron beams," *High Energy Photon and Electron Symposium*, edited by Kramer, Suntharalingam, and Zinninger (Wiley, New York, 1976).
[6]"Radiation dosimetry: electrons with initial energies between 1 and 50 MeV," ICRU Report 21 (1972).

[7] J.D. Steben, K. Ayyangar, and N. Suntharalingam, "Betatron electron beam characterization for dosimetry calculations," Fourth Intl. Conf. on Med. Physics, Ottawa (1976), submitted to Phys. Med. & Biol.

[8] J. Dutreix and A. Dutreix, "Film dosimetry of high energy electrons," Ann. N.Y. Acad. Sci. 161, 33 (1969).

[9] J.R. Cameron, N. Suntharalingam, and G.N. Kenney, *Thermoluminescent Dosimetry* (Univ. of Wisc. Press, Madison, 1968).

[10] Nordic Assoc. of Clinical Physics, Acta Radiol. Ther. Phys. Biol 11, 603 (1972).

2. Patient specific data acquistion in electron beam treatment planning

James A. Purdy, Ph.D.

Washington University School of Medicine, St. Louis, Missouri

2.1 INTRODUCTION

The task given to me for this symposium was to ascertain what type of patient information was currently being utilized in electron beam treatment planning. In addition, I want to put forth my views on what type of data will be required for future progress in this area. To accomplish this task, over 30 institutions were contacted and their electron beam treatment planning procedures were reviewed. Six aspects of electron beam treatment planning were considered, namely: 1) patient contour, 2) normal structure localization, 3) tumor localization, 4) inhomogeneities, 5) field shaping, and 6) simulation. The results of this survey are as follows.

2.2. CONTOUR

Methods for obtaining the patient's contour now being utilized or developed at those institutions surveyed included the use of 1) a solder wire or contouring ruler, 2) a plaster strip or other type of casting, 3) various multiple rods or needle devices, 4) pantographic devices, including ultrasound units, 5) electro-optical range finder devices, 6) stereophotography, and 7) computed tomography. For completeness, a review of the literature was also conducted, and an extensive set of reference on contouring devices is included.[1-10] All of the first five methods offer certain limitations including inaccuracies, being somewhat time consuming, and/or being dependent on the skill of the operator. The sixth method (shown in Fig. 2.1) is under development by several groups including our own at the Mallinckrodt Institute of Radiology and shows some promise.[11] As illustrated in Fig. 2.2, a computed tomography (CT) scan also provides patient contour information, and many of the institutions surveyed either were using CT scans for treatment planning or were planning to in the near future. However, some scanners, including the EMI scanner with which I am most familiar, use curved couches to support the patient. This curvature can cause a distortion in the patient's topography. In addition, this scanner requires the use of bolus-like material placed on the patient's surface which results in a surface compression. Another more subtle point concerns patient positioning. For example, arm position effects both the contour and internal organ positions for such treatments as chest wall irradiation. The small scan ring utilized on some CT scanners may not permit the duplication of such a therapy setup.

While all of the methods listed above for obtaining the patient's contour were being utilized for external beam (photon) treatment plan-

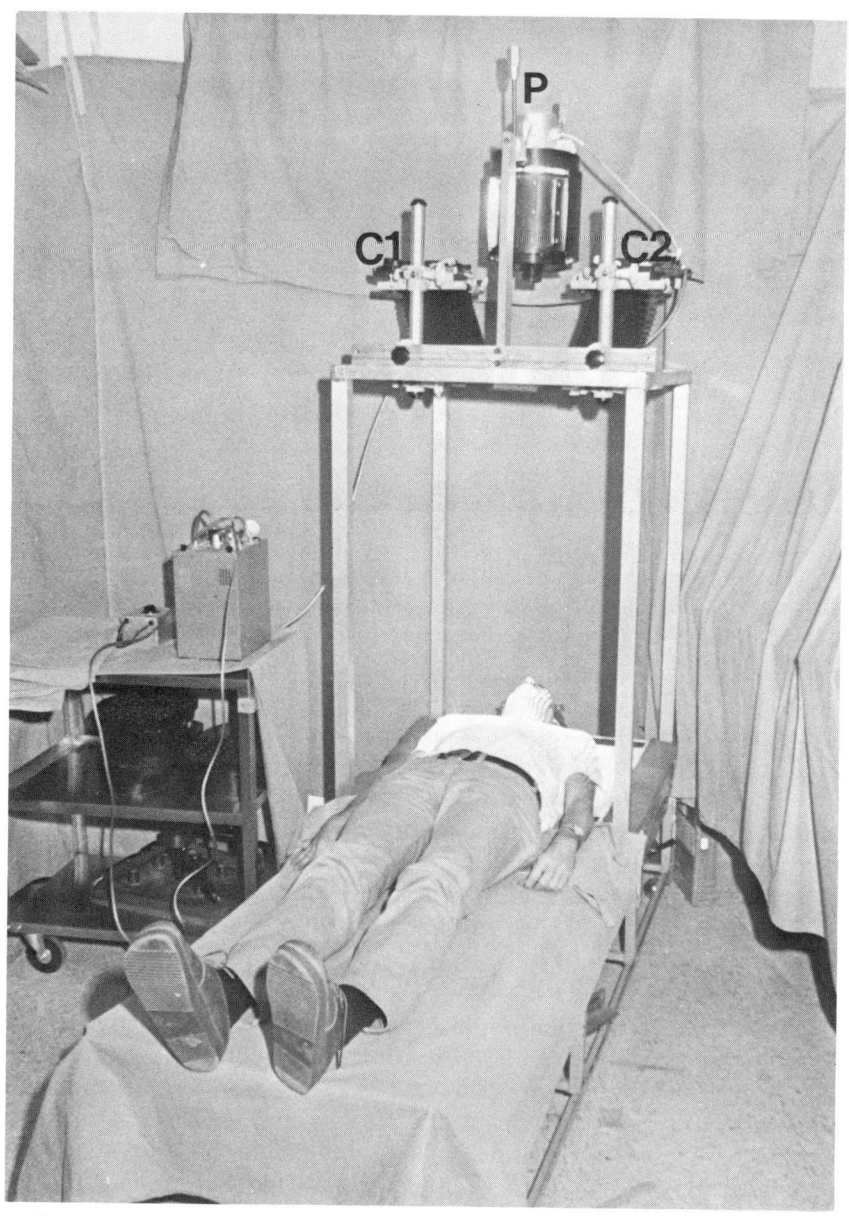

FIG 2.1A. Photograph of subject in position to be stereophotographed. C1 and C2 are two cameras, P is a flash projector. The distance from the subject to the lens plane is approximately 1 meter.

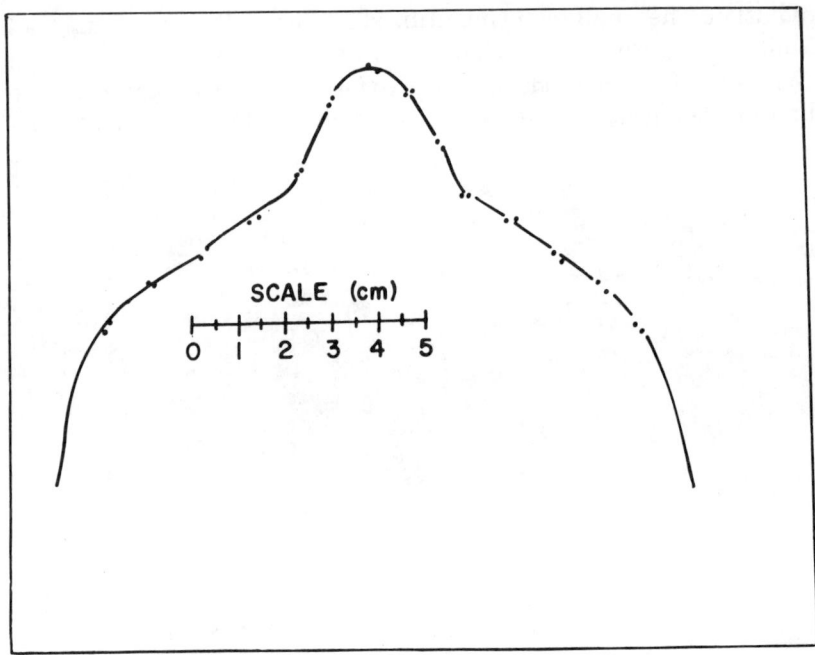

FIG 2.1B. A comparison of a contour of a Rando[R] phantom section obtained using stereophotography system and with a direct tracing (—). The agreement is excellent, with an RMS deviation of 0.7 mm. When fully developed the stereophotographic system should allow one to obtain complete and precise surface contours with no discomfort to or distortion of the patient.

ning, only 12 of the 30 institutions surveyed required such information for their electron beam treatment planning procedures. The majority of the institutions were calculating the dose along the central axis at the depth of the 80th percentile and/or relying on TLD or film dosimetry for complex cases such as abutting fields and thus did not require a full representation of the contour.

2.3. NORMAL STRUCTURE LOCALIZATION

The various techniques for delineation of cross-sectional anatomy of patients scheduled for radiation therapy used at the institutions surveyed and cited in the literature include: 1) orthogonal radiographs, 2) conventional transverse tomography, 3) standard anatomical displays available from pictorial atlases, 4) ultrasound scanning, and 5) computed tomography.[7,12-20] Again, each method suffers from various limitations including inaccuracies, the need for special viewing devices, being time consuming, etc. Method 3) is no more than a crude approximation because it offers no compensation for anatomic variations among patients. The usefulness of ultrasound is limited in the thorax, the very region where dose perturbations caused by inhomogeneities may be the greatest. Several investigators have applied the method of ultrasound to the determination of the thickness of the chest wall after mastectomy for electron beam irradiation as shown in Fig. 2.3.[12,13,16-19] Others have stressed the use of conventional transverse tomography for organ localization.[15] Again artifacts may be present in both methods, and experience is required for interpretation of the images obtained. Some investigators have combined the advantages of ultrasound and transverse tomography utilizing the ultrasound transducer to obtain the patient contour and transverse tomography to provide information on the internal structures.[7] It should be noted, though, that while both transverse tomography and ultrasonic methods give direct information about the position and shape of internal structures, they give only vague information about their nature.

There is no question in my mind that computed tomography is superior to all the methods mentioned above for normal structural localization. Figure 2.2 clearly shows the location of the lungs and spinal cord and the thickness of the chest wall. However, the problems mentioned above relating to contour distortion, i.e., 1) curved couch, 2) bolus on the

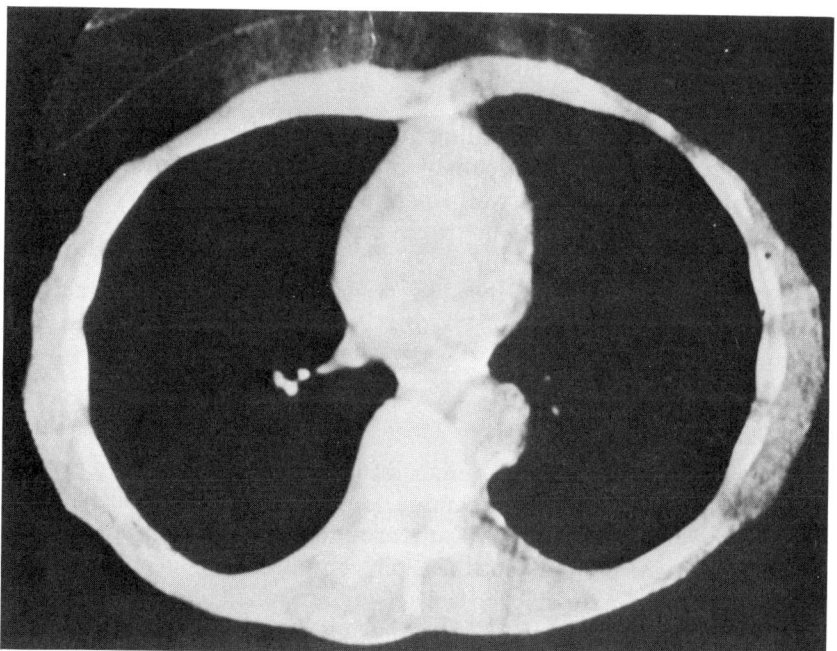

FIG. 2.2. Computed tomography scan illustrating delineation of body contour of patient, lung contour and spinal cord. In addition it appears likely that electron density information of the various tissues can be obtained from the accompanying CT numbers.

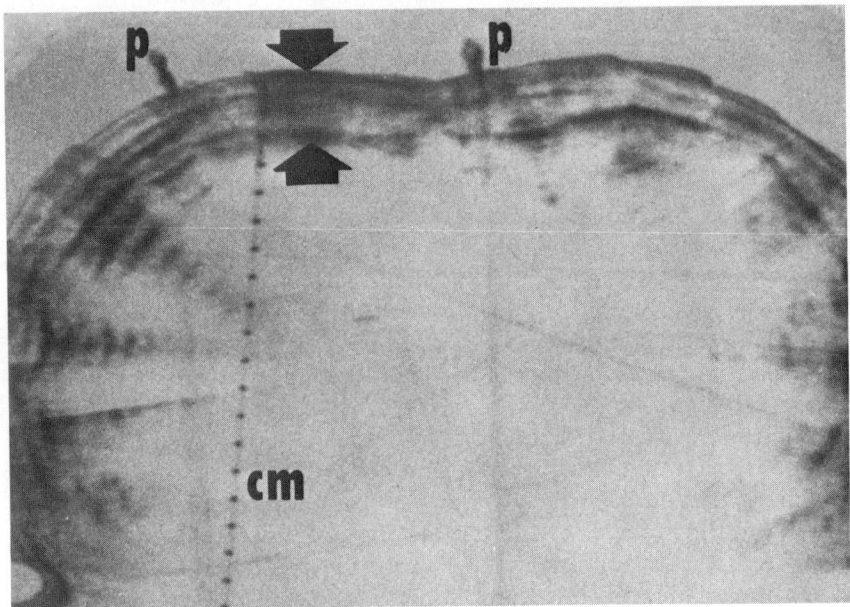

FIG 2.3. Ultrasound determination of chest wall thickness for planning electron beam therapy post mastectomy. Upper arrow, skin surface; lower arrow, plural surface. Dots indicate 1 cm distance. Reproduced with permission of Brascho (personal communication).

surface, and 3) patient positioning are pertinent here also and must be eliminated if CT is to significantly improve the patient model in therapy treatment planning. Twelve of the institutions surveyed utilized ultrasound or transverse tomography routinely in their electron beam treatment planning procedures. Several institutions[7] also indicated that CT scans were being utilized for treatment planning, e.g., in determining chest wall thickness, lung localization, and spinal cord localization.

2.4. TUMOR LOCALIZATION

The survey indicated that tumor localization at this time is primarily determined by 1) clinical examination, 2) orthogonal radiographs, 3) nuclear medicine procedures, 4) ultrasound, and/or 5) computed tomography. As regards tumor localization, computed tomography appears at first glance to be superior to the above-mentioned techniques. Recall, though, that the CT scan is representing differences in attenuation characteristics of the various regions. Tumor localization for any radiographic technique depends on the differences of attenuation between tumor and adjacent normal tissues. If no differential attenuation is present, the contrast between the tumor and adjacent soft tissues is not present. This is true for CT scanning as well as conventional radiographic techniques. Specifically, at our institution we have been unable at times to differentiate mediastinum tumor from normal structures in the mediastinum, and we have been unable to differentiate some retroperitoneal tumors from adjacent normal structures because of the blending of tissue boundaries and the lack of differential attenuation. In terms of tumor localization, then, it is likely that the CT scan will be only complimentary to the methods listed above.

2.5. INHOMOGENEITIES

Accounting for the effects of inhomogeneities in electron beam treatment planning can be quite complicated. Electron scattering depends strongly on the atomic number of the medium, while the absorption of electrons is primarily determined by the number of electrons per gram. As the dose at any point is the sum of the dose from the primary beam plus the dose from the scattered electrons, the absorbed dose will be different in materials of different density. Several clinical situations are encountered regularly, for example, when the dose is required in soft tissue beyond an inhomogeneity such as bone or lung, or when the dose is required in the inhomogeneity itself.

Several methods have been reported to correct the dose distribution for the presence of lung and bone, namely, the absorption equivalent thickness (AET) method of Laughlin, the electron absorption coefficient of Dahler et al., the coefficient of the equivalent thickness (CET) of Almond et al., and the modified absorption coefficient (MAC) method of Bagne.[21-24] The description of these methods will be discussed by other speakers at the symposium, but let me say that all of these methods require knowledge of the electron density of the tissues involved. Those institutions employing these methods generally use predetermined values rather than attempting to determine the electron densities appropriate for a particular patient. Recent developments in computed tomography may change this, though, as several investigators have shown that the CT number can be empirically related to the electron density of the tissue in question.[25-27] It is likely then that a CT method of density determination will be used more frequently in the future.

It should be pointed out that only seven of the institutions surveyed indicated that they routinely account for inhomogeneities in their electron beam treatment planning procedures.

2.6. FIELD SHAPING

The treatment of tumors with electron beams sometimes requires considerable beam shaping. Lead sheets cut in the shape of the treatment portal are often used in conjunction with the circular or rectangu-

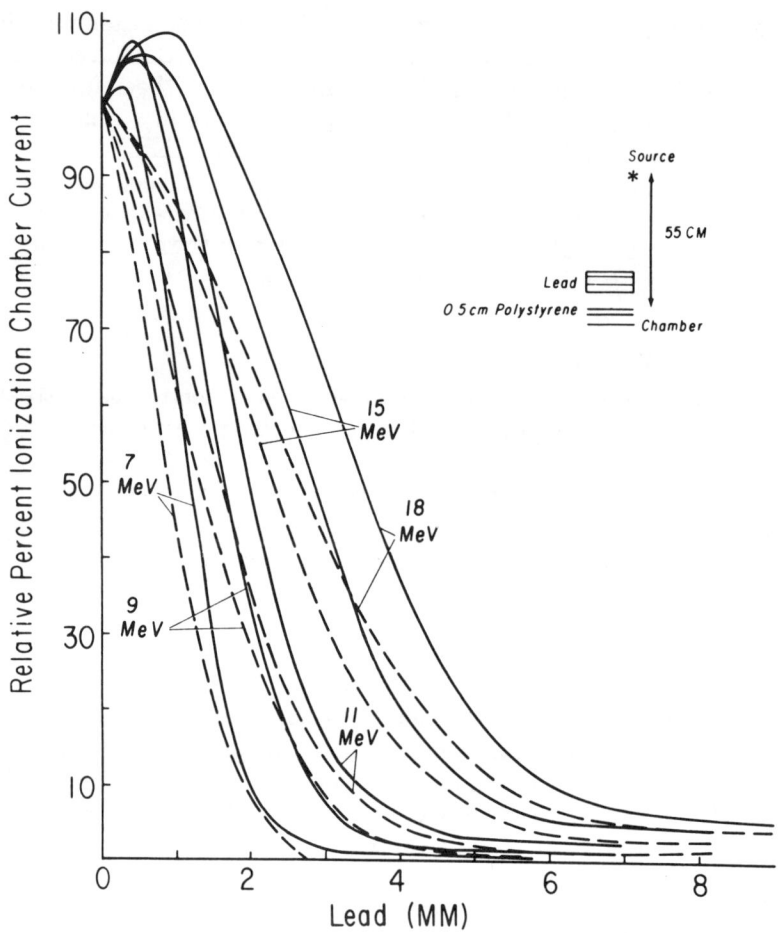

FIG 2.4. Relative percent ionization curve of 7, 9, 11, 15, and 18 MeV electrons from a Siemens Betatron incident on lead for a 6×6 cm cone (6.3×6.3 cm effective field size, dashed line) and a 10×10 cm cone (10.5×10.5 cm effective field size, solid line) at 55 cm SSD, measured to the 0.5 cm thick polystyrene on top of the ionization chamber (see insert). Accelerator energies are shown. Reproduced from Giarratano et al.[28]

lar cones generally available with electron beam units. The lead mask defines the treatment area while shielding the surrounding normal tissue or critical organ. Additional shielding may be required to protect internal structures beyond the treatment volume such as found in the treatment of the buccal mucosa.

The thickness of lead shielding is usually chosen to reduce the dose by 95% to 98% relative to the open beam. However, there are practical limits on the amount of lead that can be used, and it is important to use the minimum amount of lead to achieve the desired reduction in dose in order to minimize patient discomfort. Giarratano et al. measured the relative percent dose reduction by lead for electrons having accelerated energies between 7 and 28 MeV.[28] Results of those measurements are shown in Fig. 2.4. Note that very thin layers of lead may enhance rather than reduce the surface dose depending on the field size.

While the above method is acceptable, it does have drawbacks which may include discomfort to the patient when large pieces of lead must be used. A method that is widely employed by those institutions having accelerators utilizing cones, consists of fabricating shields from Cerrobend which may be attached to the end of the treatment cone by screwing two or more small bolts into existing holes or by adding a slide rail to the collimator.

Another method that is currently being utilized is the use of lead sheets in conjunction with a cast (either a plaster cast or "light cast"). Sheet lead is pressed formed to cover the cast. The lead is attached to the cast with staples. An example of this type of field shaping is shown in Fig. 2.5.

One institution utilizes a scanning electron beam.[29] This novel approach to field shaping allows arbitrary field sizes and shapes up to 21 cm × 21 cm to be accomplished by using magnetic fields to guide a "pencil" beam (approximately $\frac{1}{2}$ cm in diameter) over the tumor area of a patient.

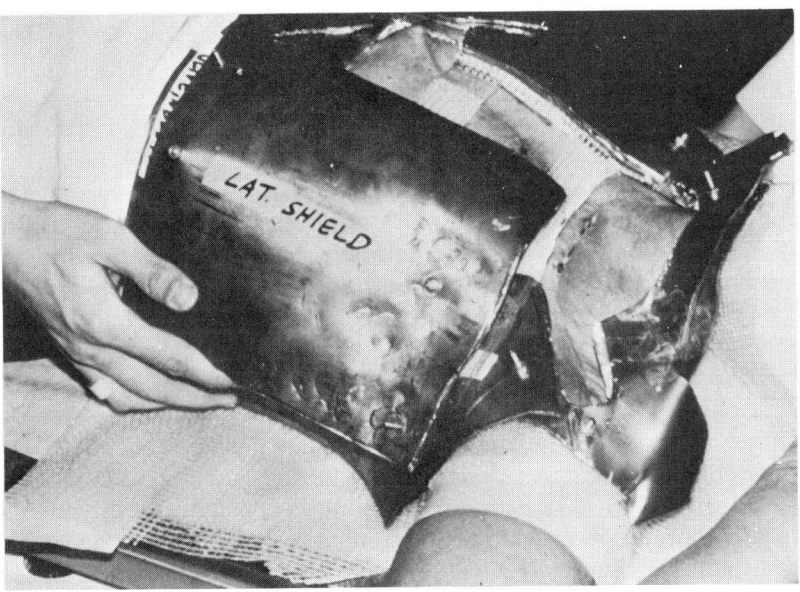

FIG. 2.5. Photograph of field shaping device currently used at the Mallinckrodt Institute of Radiology for electron beam field shaping. Sheet lead is pressed formed to cover the casting material. The lead is attached to the cast with staples. The device then serves to reposition the patient and define the target area while shielding the non-involved region.

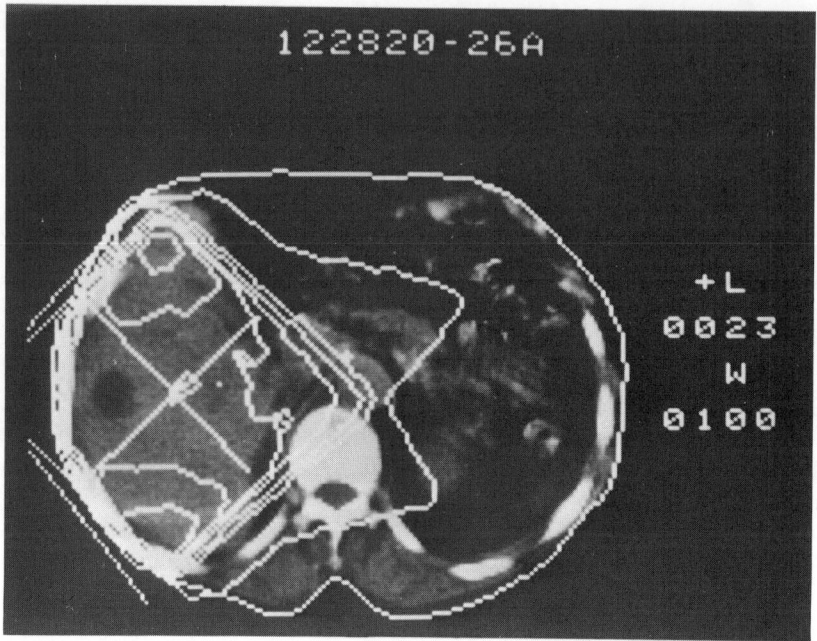

FIG 2.6. Cathode ray display of CT scan showing superimposed radiation beams and subsequent isodoses.

2.7. SIMULATION

In this section of the paper, I would like to generalize my impression of the treatment planning procedure based on the results of the survey, beginning with simulation. In general, radiographs are used to determine the volume to be irradiated. The physician then marks on the patient's skin the portal to be irradiated. The patient is placed in the treatment position and the portal outline is obtained (usually by placing a transparency over the patient and tracing the portal outline with a marking pen). This template is then used to construct a mask either from Cerrobend or lead sheets.

For clinical situations in which a single field is to be used, no field shaping is to be employed, and no inhomogeneities are felt to be of concern, a calculation along the central axis at the depth of d_{80} or d_{90} is in general the extent of treatment planning performed. As the clinical situation became more complex, i.e., abutting fields, shaped portals, large contour irregularities, etc., the majority of the institutions surveyed relied mainly on TLD or film dosimetry rather than isodose computation, expressing little confidence in presently available computer software. In fact, most groups felt, and I fully agree, that there was an urgent need for improved electron beam treatment planning software, and I am confident later speakers will address that problem.

Future advances in electron beam treatment planning computations will require patient data specifying organ location, shape, and electron density. The requirements for an improved patient model appear to be provided by recent advances in computed tomography. Figure 2.6 shows a CT scan displayed on a cathode ray viewing screen. Appropriate software has been developed which allows one to superimpose radiation beams on the transverse section and display the subsequent isodoses. Much work still remains, though, in developing algorithms to fully utilize the structure location and density information shown in this figure.

2.8. REFERENCES

[1] Paul W. Scanlon, "A simple method for the mensuration of body contour," Radiol. **74**, 968–70 (1960).

[2] Kai Setala, "Automatic body-contouring unit for dose planning in radiotherapy," Acta Radiologica (Therapy) **3**, 269–80 (1965).

[3] Hector C. Clarke, "A contouring device for use in radiation treatment planning," Br. J. Radiol. **42**, 858–60 (1969).

[4] L.H. Lanzl, T.J. Ahrens, M. Rozenfeld, and L. Bess, "An automatic patient-contour measuring apparatus," Am. J. Roent. **108**, 162–71 (1970).

[5] C.B. Clayton and D.J. Thompson, "An optical apparatus for reproducing surface outlines of body cross-sections," Br. J. Radiol. **43**, 489–92 (1970).

[6] J. Legal, A.F. Holloway, and K. Breitman, "A contour plotting device," Am. J. Roent. **111**, 182–83 (1971).

[7] Michael Hughes and Edward S. Sternick, "A method for obtaining body contours using an ultrasonic scanner and transverse tomograms," Med. Phys. **1**, 72–73 (1974).

[8] S.C. Lillicrap and J. Milan, "A device for the automatic recording of patient outlines on the treatment simulator," Phys. Med. Biol. **20**, 627–31 (1975).

[9] A.M. Doolittle, Jr., M. Phil, L.B. Berman, G. Vogel, A.G. Agostinelli, Carolyn Skomro, and R.J. Schultz, "An electronic patient-contouring device," Br. J. Radiol. **50**, 135–38 (1977).

[10] Larry D. Simpson and Radhe Mohan, "Computer-compatible patient plotter," Med. Phys. **4**, 215–19 (1977).

[11] D.E. Velkley, G.D. Oliver, Jr., J. Wetzel, and A.E. Vorlage, "Stereo-photogrammetric determination of patient surface geometry," Radiation Oncology Research and Treatment Center Report, 1975–76.

[12] Edward H. Smith and Hans H. Holm, "Ultrasonic scanning in radiotherapy treatment planning," Radiol **96**, 433–35 (1970).

[13] William N. Cohen and A. Curtis Hass, "The application of B-scan ultrasound in the planning of radiation therapy treatment ports," Am. J. Roent. **111**, 184–88 (1971).

[14] F.T. Farmer and Margaret P. Collins, "A new approach to the determination of anatomical cross-sections of the body by Compton scattering of gamma-rays," Phys. Med. Biol. **16**, 577–86 (1971).

[15] L.D. Simpson, R.C. Fleischman, F. Chu, and J.S. Laughlin, "Physical methods for determination of the nature and the extent of internal anatomy for radiation treatment design and computation," *Proceedings of the 13th International Congress of Radiology, Madrid, 15–20 October 1973* (Excerpta Medica, Amsterdam, Am. Elsevier, New York, 1974), p. 565–74.

[16] John Eule, Jr., Fred Bockenstedt, and Emanuel Salzman, "Diagnostic ultrasound scanning: A valuable aid in radiation therapy planning," Am. J. Roent. **117**, 139–45 (1973).

[17] James M. Slater, Ivan R. Neilsen, William T. Chu, Ernest N. Carlsen, and Jere E. Chrispens, "Radiotherapy treatment planning using ultrasound-sonic graph pen-computer system," Cancer **34**, 96–99 (1974).

[18] K.E. Ekstrand, R.L. Dixon, D.D. Blake, and M. Raben, "The calculation of dose distribution for chest wall irradiation using B-mode ultrasonography," Radiol. **111**, 185–87 (1974).

[19] Donn J. Brascho, "Computerized radiation treatment planning with ultrasound," Am. J. Roent. **120**, 213–23 (1974).

[20] G.A. Thieme, W.R. Hendee, G.S. Ibbott, P.L. Carson, and D.L. Kirch, "Cross-sectional anatomic images by gamma ray transmission scanning," Acta Radiol. Ther. Phys. Biol. **14**, 81–112 (1975).

[21] J.S. Laughlin, "High energy electron treatment planning for inhomogeneities," Br. J. Radiol. **38**, 143 (1965).

[22] A. Dahler, A.S. Baker, and J.S. Laughlin "Comprehensive electron beam treatment planning," Ann. N.Y. Acad. Sci. **161**, 198–213 (1969).

[23] P.R. Almond, A.E. Wright, and M.L.M. Boone, "High energy electron dose perturbations in regions of tissue heterogeneity," Radiol. **88**, 1146–53 (1967).

[24] Farideh Bagne, "Electron beam treatment-planning system," Med. Phys. **3**, 31–38 (1976).

[25] P.S. Rao and E.C. Gregg, "Attenuation of monoenergetic gamma rays in tissues," Am. J. Roent. **123**, 631–37 (1975).

[26] Michael E. Phelps,, Mokhtar H. Gado, and Edward J. Hoffman, "Correlation of effective atomic number and electron density with attenuation coefficients measured with polychromatic x rays," Radiol. **117**, 585–88 (1975).

[27] R.A. Rutherford, B.R. Pullan, and I. Isherwood, "Measurement of effective atomic number and electron density using an EMI scanner," Neuroradiology **11**, 15–21 (1976).

[28] Joseph C. Giarratano, Robert J. Duerkes, and Peter R. Almond, "Lead shielding thickness for dose reduction of 7- to 28-MeV electrons," Med. Phys. **2**, 336–37 (1975).

[29] L.H. Lanzl, "Electron pencil beam scanning and its application in radiation therapy," Front. Radiation Ther. Onc. **2**, 55–66 (1968).

3. Treatment planning consideration

Jacques Ovadia, Ph.D.
Michael Reese Hospital and Medical Center, Chicago, Illinois

3.1 INTRODUCTION

I want to discuss certain aspects of treatment planning and treatment execution with high energy electrons, and shall limit the discussion chiefly to electrons in the energy range of about 10–35 MeV, with which I have most experience. Electrons in this energy range have been used for about 20 years.[1-5]

The proceedings of the first symposium on High Energy Electrons in 1964[1] identified most of the points that will be briefly alluded to here, namely:

Effect of the collimation system on the depth dose distribution.

Field shaping at the patient location to protect sensitive organs, such as the eye.

Multiple field techniques.

Isodose measurements in inhomogeneous matter.

Treatment planning, taking inhomogeneities into account.

Treatment plans in breast cancer therapy.

There were also reports on integral dose and on the dose to the patient from neutrons contaminating the electron beam, which were brought up again over 10 years later in the Philadelphia Symposium[5] and which I will discuss briefly.

I want now to say a few words about each of the above topics, plus two additional ones: computers and specifications of tumor dose.

3.2 CENTRAL AXIS DEPTH DOSES

It is well known that high energy electrons lose about 2 MeV per centimeter of tissue, leading to a finite range (Fig. 3.1). These particular data were obtained about 20 years ago, and the exact shape of the central axis depth dose is significantly influenced by the particular scattering and collimation system we were then using. It has also been clear for a long time, from measurements by Larry Lanzl at the University of Chicago Linear Accelerator which uses a scanned electron beam like the Sagittaire at M.D. Anderson, that, for a given electron energy, the penetration of a scanned beam is greater than that of a scattered beam.

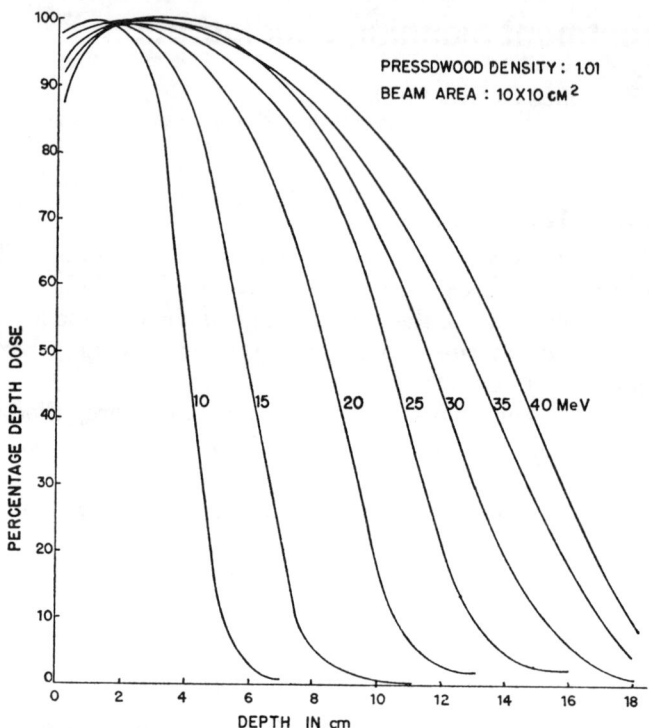

FIG. 3.1. Central axis depth dose curves in Pressdwood for a 10×10 cm² beam of electrons of various energies.

I have found the following rule of thumb convenient: *the depth in centimeters where electrons deliver a useful dose, i.e., to the 70%–80% isodose level, is equal to about one third of the electron energy in MeV.* Thus a 15 MeV beam is useful to a depth of about 5 cm. This makes electrons ideal for the treatment of superficial tumors in the skin or the chest wall, where the electron energy is selected to provide a range that encompasses the target depth but spares underlying tissue such as the lung. This will be discussed later by Dr. Bagne (Sec. 6, pp. 80–96).

I would like to make one small point concerning chest wall irradiation, including the sternum or ribs. There were early publications, mainly from the M.D. Anderson Hospital, where great attention was focused on the need for the accurate determination of the chest wall thickness and for corrections to allow for the higher bone density, as requirement for treatment plans to be clinically applicable. I believe the real problem was that the electron energy was marginal in some cases, and with the sharp fall-off of the central axis depth dose due to the higher bone density, the results were often disastrous, in terms of delivering the desired dose. The obvious solution is that, when in doubt,

3. Treatment planning considerations   35

use a higher electron energy to make sure that the target volume in the chest wall is well within the high dose range. Developments in ultrasound are additional factors which reduce the significance of this problem, as of this date.

One last comment on penetration: I am perfectly willing to go along with the opinions expressed at the symposium in Philadelphia[5] that

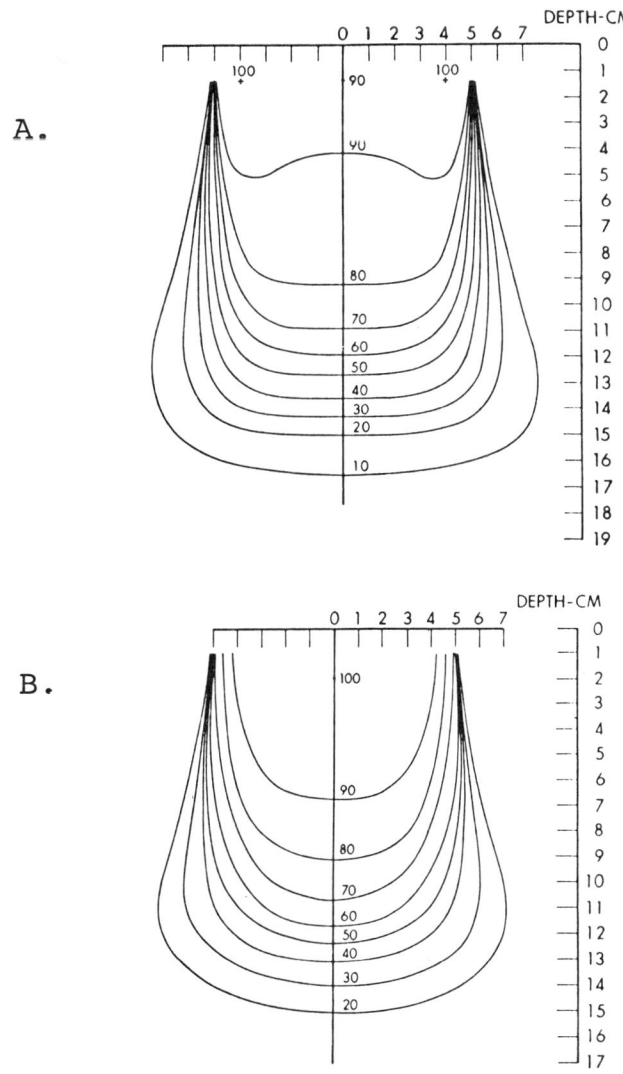

FIG. 3.2. Isodose curves obtained with 10 cm wide cones:
  A. Isodose curves (35 MeV electrons) from the modified 10×10 cm cone.
  B. Isodose curves (35 MeV electrons) obtained with a standard 10×8 cm Lucite cone. (From Schultz[2]).

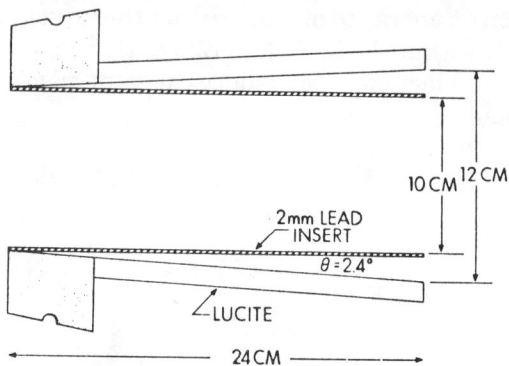

FIG. 3.3. A standard 12 × 12 cm Lucite treatment cone modified by lead inserts to form a 10 × 12 cm field. (From Schultz[6]).

electrons with maximum energy of about 25 MeV are in general adequate to treat most tumors for which they are useful, provided one treats from one side only. However, when one treats centrally located tumors by parallel opposed fields, higher electron energies lead to a better distribution.

3.3 SINGLE FIELD ISODOSE DISTRIBUTIONS

Above approximately 10–15 MeV, one can no longer plan treatments from central axis depth-dose data, but must use isodose distributions. The basic reason is scatter, both in the patient and in the collimating structure. This is illustrated in Fig. 3.2. The isodose curves always exhibit curvature, but the actual shape can be modified and sometimes improved by the proper choice of material and geometry of the collimator. The improvement from the 2 mm of lead insert is dramatic (Fig. 3.3). In particular, I do not believe that Lucite or other low atomic number plastics are suitable collimator material. We all know that the use of Lucite reduces Bremsstrahlung production but this is not the whole story in view of other problems: Lucite collimators have thick walls, are heavy, awkward to handle, and often interfere with execution of the treatment, particularly in the head and neck region because the shoulder gets in the way. In my opinion, aluminum, with about three times the density is better, or better still is brass, with a density of about 8 to 9 and moderate atomic number. This is pure engineering physics where one must understand the physics and make some judicious trade-offs.

The best design of fixed collimators I am familiar with is the one used currently in the Brown-Boveri Asklepitron, where the primary collimation is done in the therapy head, and lightweight applicators trim the beam near the patient. The Sagittaire has a variable collimator with conceptually similar design, where the primary collimation is done at some distance from the patient, with additional trimming and field definition near the patient. This is illustrated in Figs. 3.4 and 3.5.

The use of isodose distributions, even for relatively low energy electrons, gives some insight into what occurs when the beam strikes a large and curved surface, such as the chest wall. Essentially, the isodoses follow the curvature, as is illustrated in Fig. 3.6.

One last point about single field treatment planning. It is often desirable to shape the field, or to block part of it, to shield structures such as the lens of the eye. As far as I am aware, the most common method is to use lead next to the patient's skin, with a thickness determined by the electron energy, but usually of the order of $\frac{1}{4}$ inch. A detailed study by Almond appears in Ref. 4.

3.4 ADJACENT FIELDS

It is often useful to treat patients with two adjacent fields, either of the same or of different energies. Phantom studies with film densitometry indicate that useful distributions can be obtained, provided some care is exercised. The situation is very different from that encountered with megavoltage x rays, where the match occurs only at one selected depth, and a more superficial region is underdosed and a deeper region is overdosed. Figure 3.7 shows the combined dose distribution from a 33 MeV and a 15 MeV electron beam, separated by a gap of 7 mm with the collimator conditions shown. The phantom was located 1.5 cm from the brass and aluminum collimator, and its edge was machined to introduce some curvature and thus decrease the lateral dose distribution gradient. In addition, a $\frac{1}{2}$ inch aluminum wedge was added to the other field, as indicated. With this arrangement, the dose near the surface in

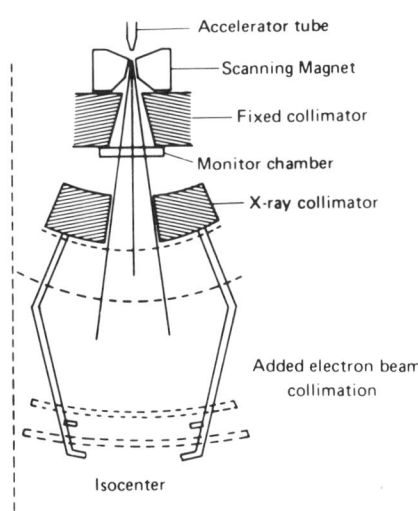

FIG. 3.4. Schematic of Sagittaire treatment head. (From Almond[4]).

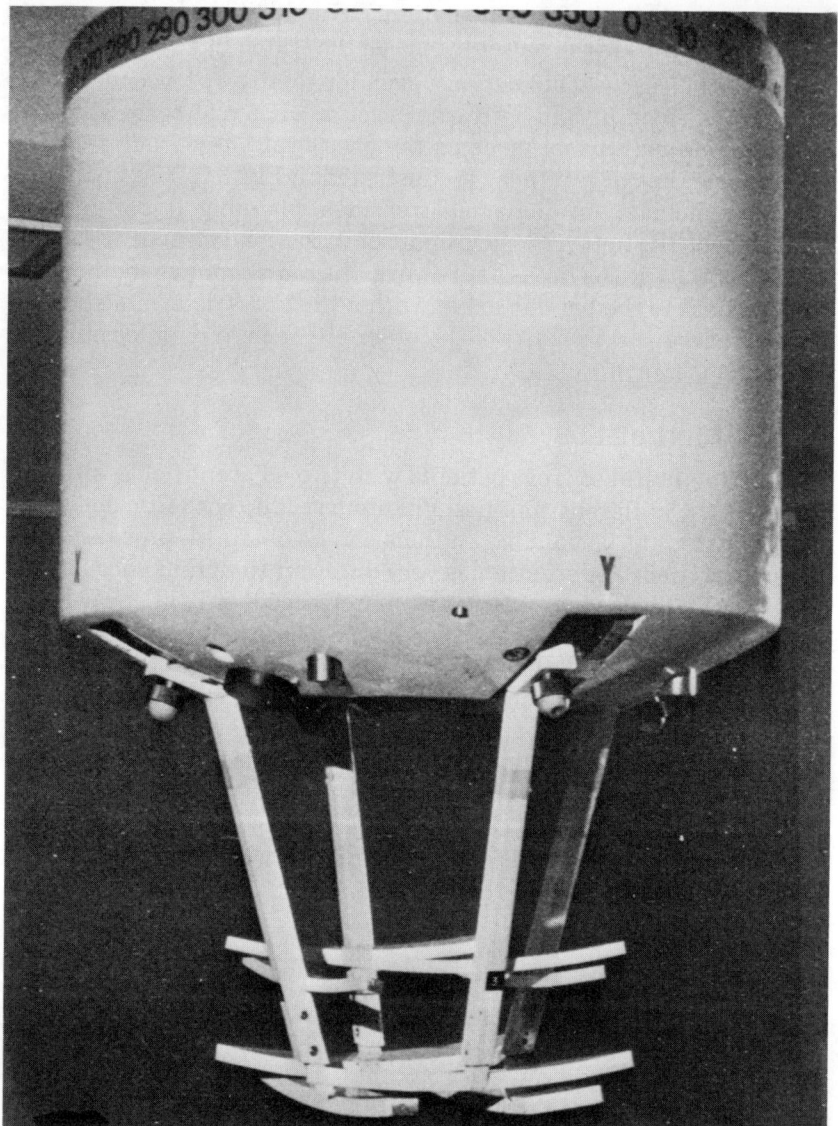

FIG. 3.5. Photograph of the Sagittaire electron-beam collimator illustrated in Fig. 4. It can be seen that the collimators are open on the side. The lower bars define the field close to the patient, while the upper bars prevent scattered electrons from reaching the patient. (From Almond[4]).

the boundary never drops below 85%, which was considered acceptable.

Figure 3.8 shows a similar study for two 33 MeV beams with gaps of 3 mm, 5mm, and 7 mm as defined above. A gap of 5 mm was adopted as providing the best match in the 50–80% isodose region.

Figure 3.9 shows a similar study published by Almond for 19 MeV electrons, with quite different results. This emphasizes the importance of carrying out a study at one's own institution with the exact conditions used for therapy, including the distance between the phantom and the collimators.

Life becomes a little more difficult when the skin surface has significant curvature, or when the fields are not parallel. Figure 3.10 shows a 1×1 cm triangular wedge used to reduce the "hot" spot at the junction of two fields.

Figure 3.11, from the work of Ward in England, shows the results of the combination of three 15 MeV electron beams, aimed at the chest wall, and each more or less perpendicular to the skin. Notice the two hot spots of 160%, where the beams overlap. This problem can be corrected by the proper selection of gap width, and by the addition of triangular wedges or other modifications of the edge of the collimator, as illustrated in Fig. 3.12 for carcinoma of the breast and in Fig. 3.13 which is our standard arrangement for post-mastectomy irradiation of axillary and supraclavicular nodes. The physicians at Michael Reese happen to like this treatment plan because the axillary, supraclavicular and mediastinal nodes are treated at the same time, with the patient in a supine position, with little risk of gaps occurring between

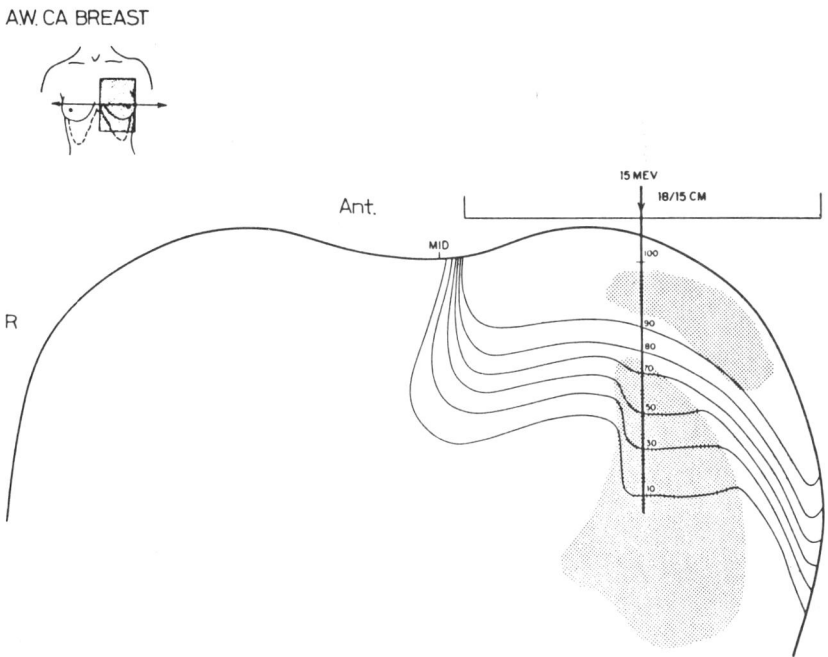

FIG. 3.6. Treatment of breast with an anterior field. (From Chu[7]).

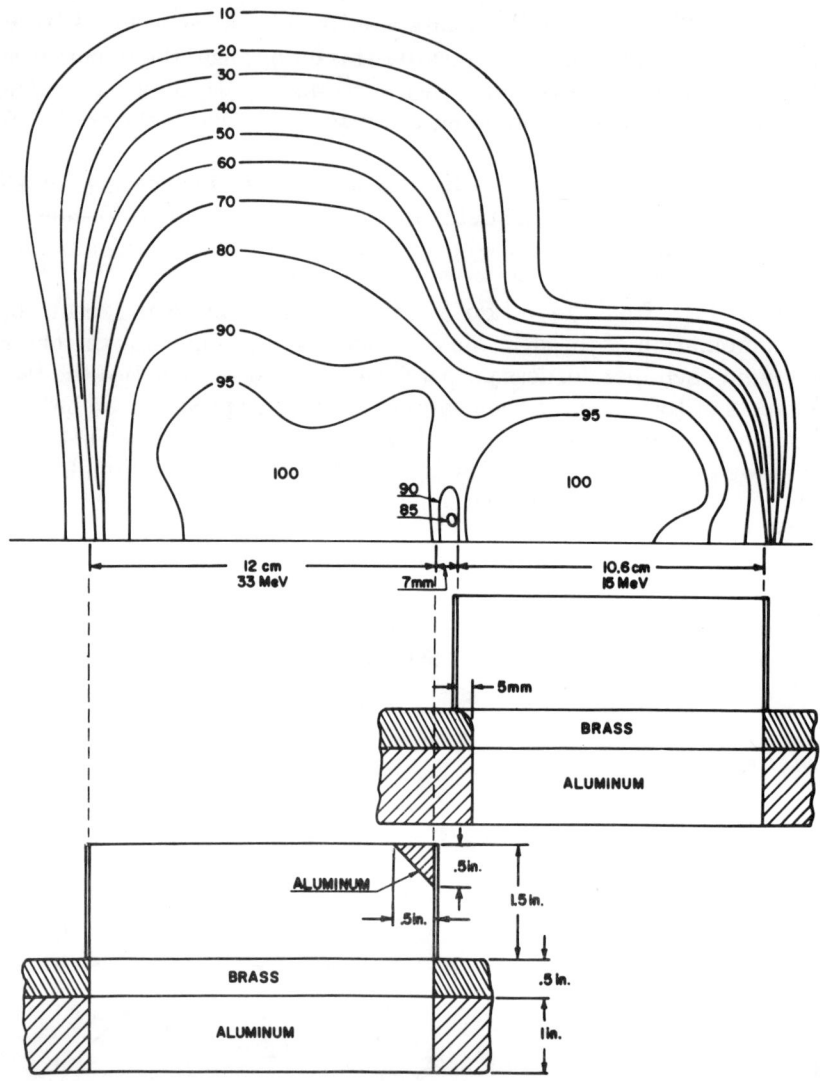

FIG. 3.7. Isodose distribution in the central plane of two adjoining fields of 33 MeV and 15 MeV electrons, and cross section of modified collimators. (From Ovadia[3]).

treated regions, as could occur in more conventional therapy. The 33 MeV electron beam, used to treat the axillary node at the 90% level, could be replaced by a megavoltage x-ray beam of an accelerator with dual electron and x-ray capabilities, but one would have to work out the interface between the electron and megavoltage x-ray beams.

3.5. ELECTRON TREATMENT PLANNING PROGRAM

This program was developed by Frank Borger in my group, runs on a PDP 11/45, and was first presented last year at the International Congress of Medical Physics in Ottawa.

The electron treatment planning program uses an empirical analytical model for the electron beam. The central axis depth-dose information is contained in a data file. The information is indexed by energy only, not by field size. This makes the model good only for fields larger than 4×6 cm. The cross beam profile is modelled in a completely analytical fashion.

The mechanics of operation of the program are very similar to the 6 MeV x-ray planning program, which has been in use for several years, i.e., interactive data entry with a small amount of information required for each field.

Although the program in its present state is restricted to fields larger than 4×6 cm, this has not proved especially hindering since most multiple field planning is done with the large fields. Figures 3.14 and 3.15 show respectively the distributions from a 10×10 cm 33 MeV electron beam, at normal incidence, and with an angle of incidence of 22°, while Fig. 3.16 shows the distribution from a 30 MeV 4×6 cm beam, the smallest field size adequately modelled.

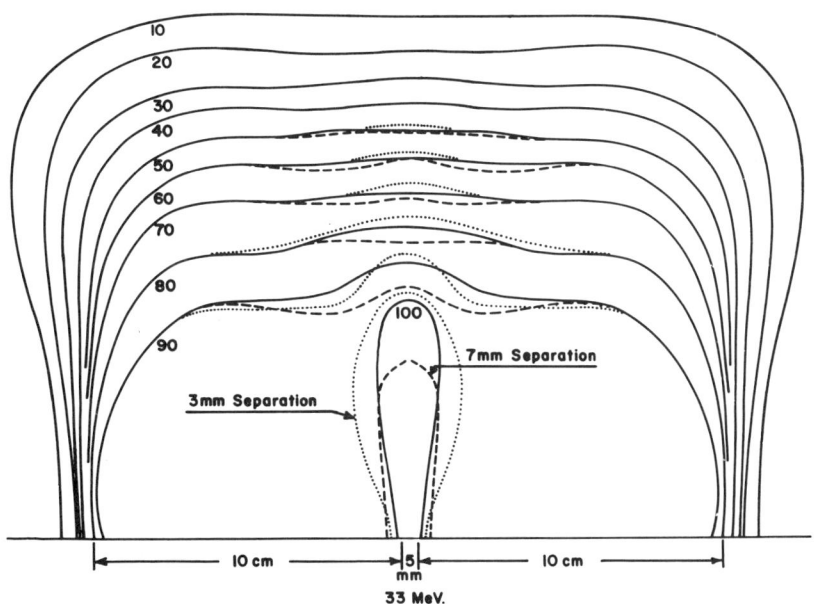

FIG. 3.8. Isodose distributions of two 10×10 cm² beams of 33 MeV electrons placed side by side, with separations of 3 mm, 5 mm, and 7 mm.

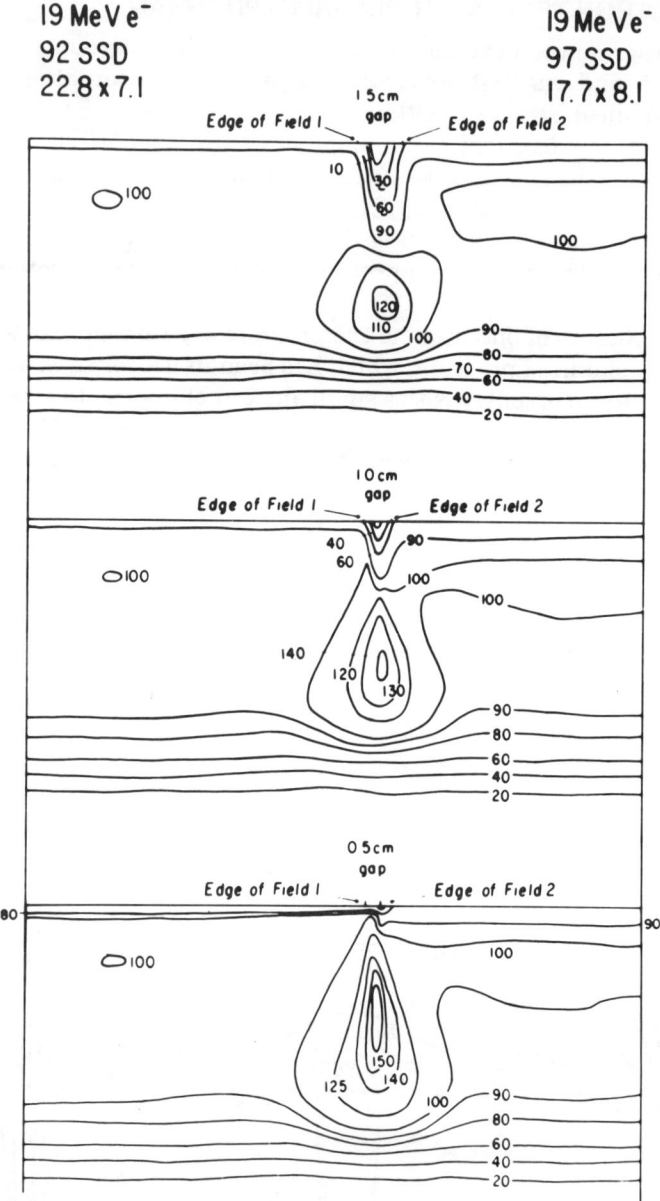

FIG. 3.9. Isodose distributions for adjoining fields with the same electron beam energy with different gap width between the fields. (From Almond[4]).

3.6. COMBINATION OF FIELDS AND TUMOR DOSE SPECIFICATION

A dose distribution for a three field irradiation of the pituitary, which appears in Fig. 3.17, illustrates two important points:

1. Whenever a centrally located tumor is irradiated, the highest en-

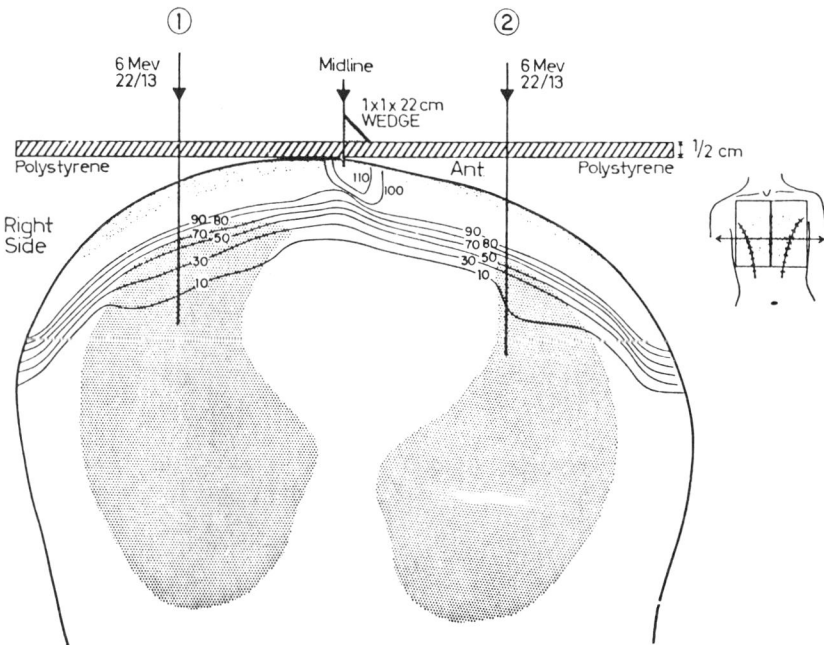

FIG. 3.10. Bilateral treatment of extensive inflammatory recurrence of breast Ca. (From Chu[8]).

ergy electron which does not produce an exit dose leads to the widest "uniformly" treated region in the center. This normally means 30–35 MeV, although in this case 25 MeV electrons were used because of the small dimension of the pituitary. I also notice that Dr. Ho from Hong

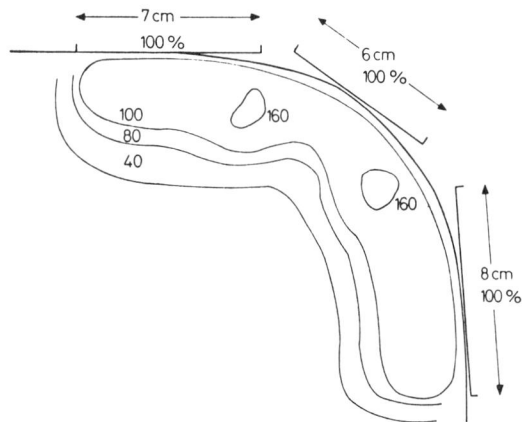

FIG. 3.11. Isodose curves for three fields of 15 MeV electrons on the chest wall. (From Ward[1]).

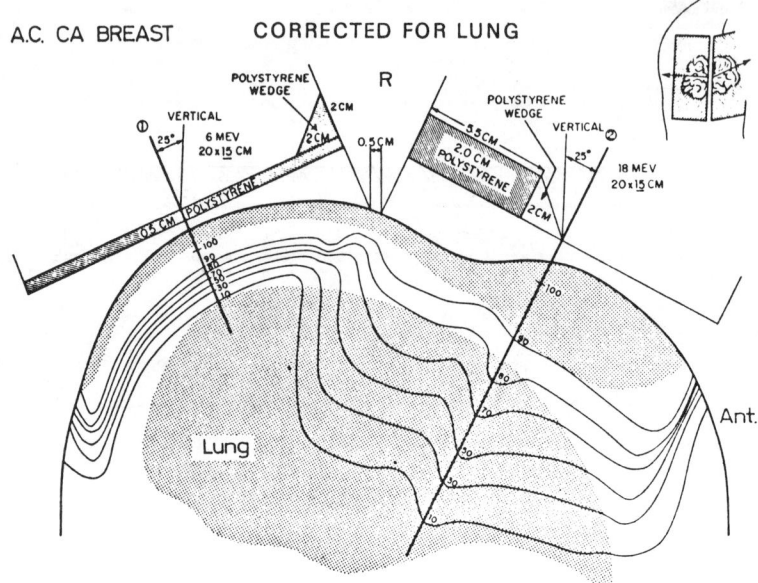

FIG. 3.12. Illustration of a plan used to treat a relatively large area of metastatic disease. Two fields are employed utilizing polystyrene absorbers to compensate for the irrgular volume of the tumor, and to avoid a "hot spot" at the interface of both fields. (From Laughlin[2]).

Kong uses 30 MeV electrons for parallel opposed fields in the head.[2]

2. Even with 25 or 30 MeV electrons, the target volume is not irradiated nearly as uniformly as could be done with megavoltage x rays. The target volume, indicated by the dotted lines, is treated to a minimum dose at the 160% isodose level, and there is one point at the center that reaches 200%. This value of 200% is the dose that would be calculated from central axis depth-dose data. Thus some people report a dose based on the 200% point, others report a dose based on the 160% level. I am convinced that an enormous amount of confusion appears in the literature of high energy electrons from this point alone. It emerged again in the discussion of Dr. Schumacher's paper in the symposium in Philadelphia[5]: whereas he reports 500 rads per weekly session in his treatment schedule for lung tumors, the dose that would be reported from the isodose at the edge of the field is 400 rads per day. This possible ambiguity should be kept in mind whenever one evaluates reports in the literature of electron therapy results and prescribed doses.

3.7. TISSUE HETEROGENEITY

The distortions of the isodose distributions produced by bone, air cavities, or lung tissue, have been known for over two decades, and a detailed study by Netteland was presented at the Montreux Symposium in 1964 (Fig. 3.18 [1]). A film study of the distortions introduced by air cavities in the nasopharynx using a cadaver phantom was reported by Dr. Ho in San Francisco in 1966.[2] Although one was aware of the problem, one did not quite know what to do with it.

Until very recently only two corrections for inhomogeneity were available, as far as I am aware: 1) those of M.D. Anderson, which I discussed, and 2) the inhomogeneity corrections described by Laughlin in great detail in the Montreux Symposium in 1964, and still in use.[1] The Memorial correction defines an empirical absorption equivalent thickness (AET), which turns out to be a function of the thickness and depth

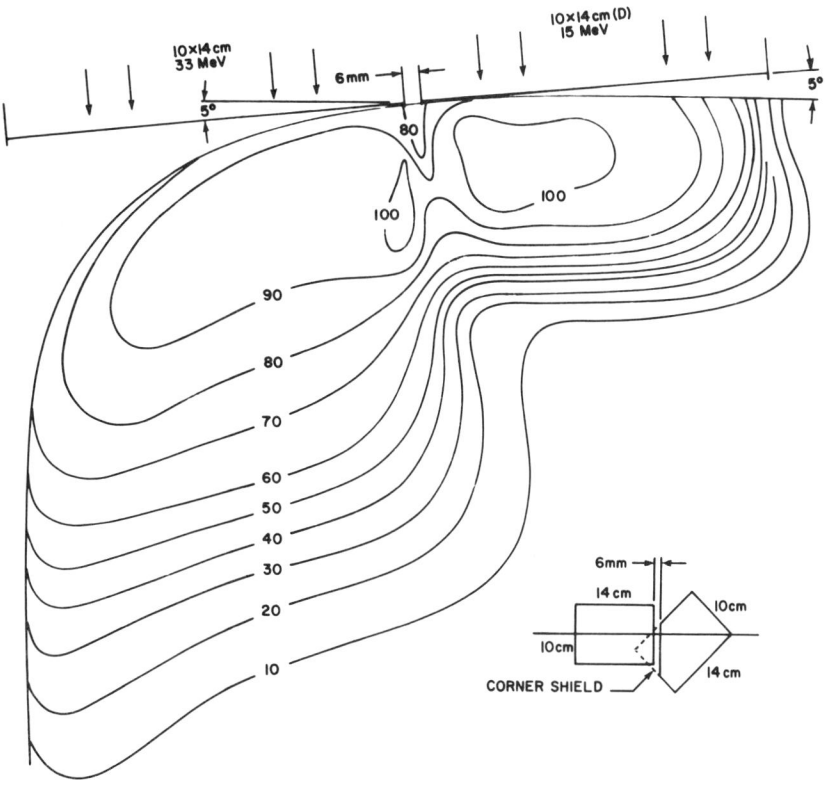

TREATMENT PLAN FOR SUPRACLAVICULAR
AND AXILLARY NODES

FIG. 3.13. Isodose distribution in treatment plan for axillary and supraclavicular nodes measured in special phantom simultating patient's contour. The relative positions of the fields appear at the bottom right corner. (From Ovadia[3]).

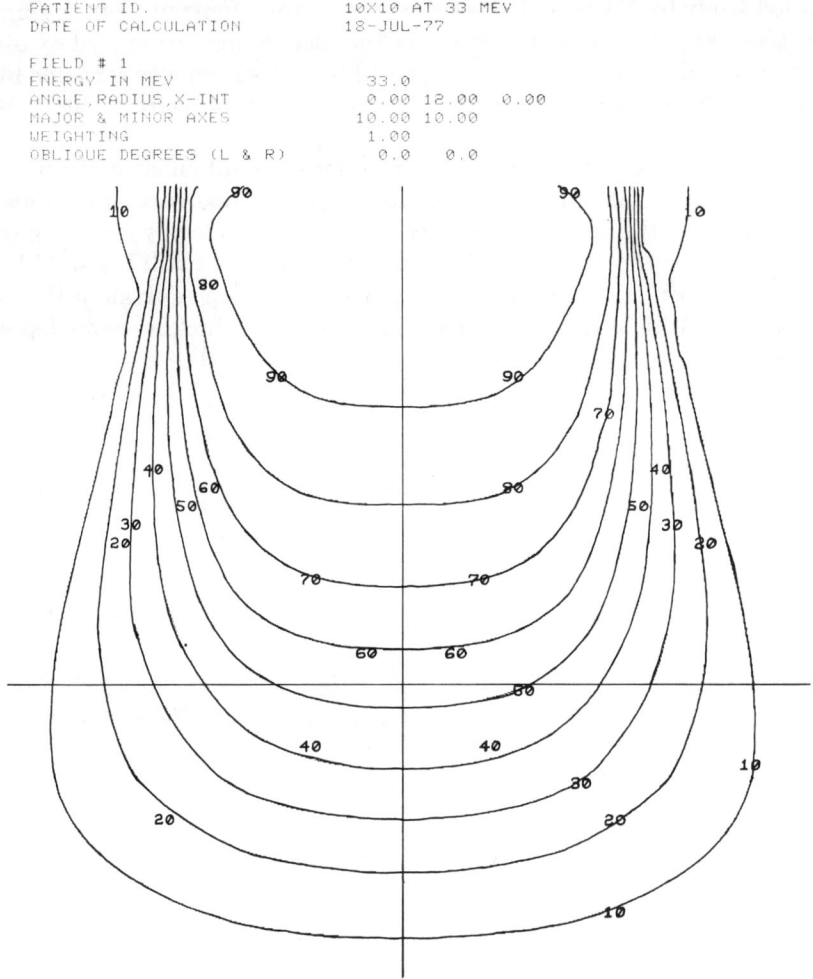

FIG. 3.14. Isodose distribution for a 10×10 cm^2 beam of 33 MeV electrons generated by the Michael Reese Computer program.

of the nontissue equivalent material such as the lung. A correction is made along rays from a virtual source for the thickness of lung tissue along that ray. Lung appears to be the only tissue where the correction may be significant. This method is reviewed in detail in Secs. 4 and 5 of this Monograph (pp. 52–79).

In my opinion, such an elaborate correction is legitimate only if one has anatomical information on the exact geometry of the inhomogeneity. Such information for a patient in treatment position could be obtained by transverse axial tomography, available in very few institutions, Memorial Hospital in New York being one of them. A treatment plan corrected for inhomogeneities appears in Fig. 3.19.

Recently Pohlit[5] described a method for calculating the modification in the isodose distribution arising from electron scatter at the edge between two media of different density, such as bone and soft tissue and this is discussed in detail by Sternick in Sec. 4 (pp. 58–60).

With the advent of CT scanning and the availability of medium sized computers, such as the PDP 11/45, I look now to the inclusion of inhomogeneity corrections in selected electron beam treatment plans.

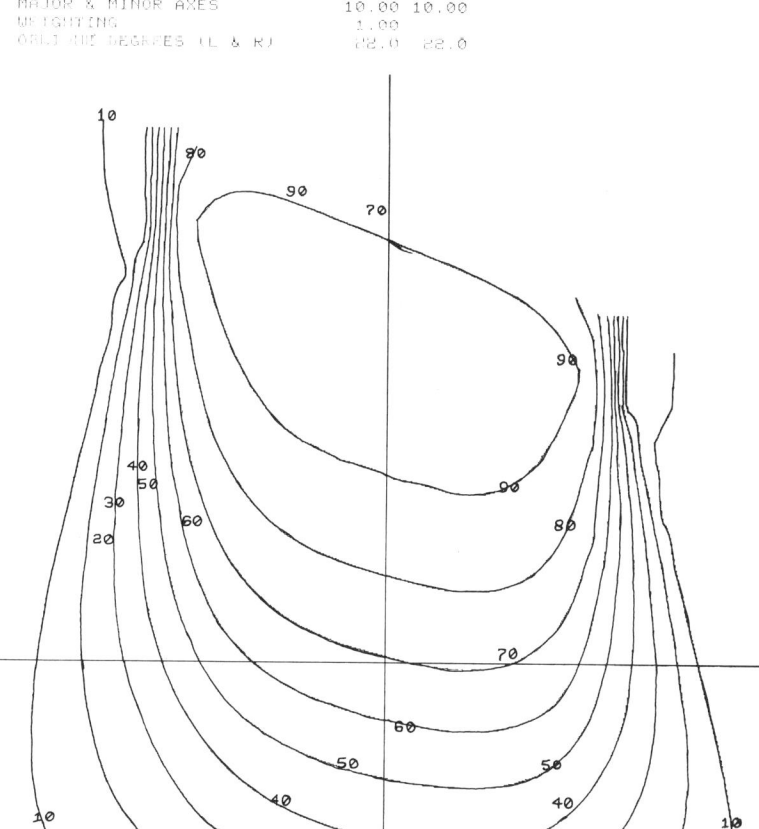

FIG. 3.15. Isodose distribution for a 10×10 cm^2 beam of 33 MeV electrons, incident at an angle of 22°, generated by the Michael Reese Computer Program.

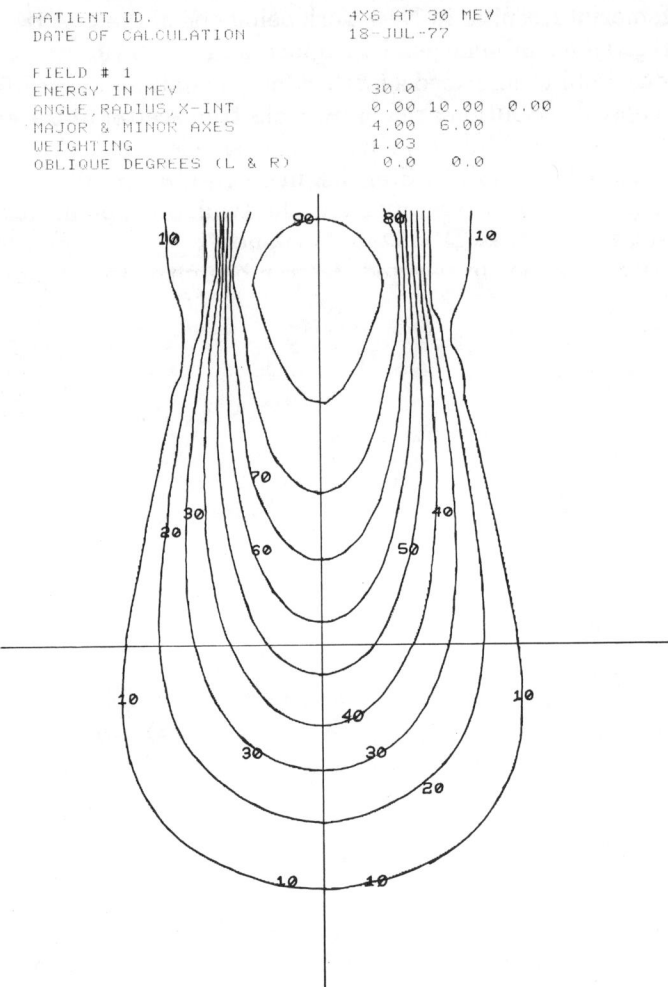

FIG. 3.16. Isodose distribution for a 4×6 cm^2 beam of 30 MeV electrons.

3.8. CONTAMINANT NEUTRON DOSE

The first study of neutron dose from a 22 MeV betatron was published by Laughlin over 25 years ago, and I had a small note in Radiology on this subject in 1956, over 20 years ago. Now the Bureau of Radiological Health is rediscovering the wheel and is engaged in a major program with computer analysis of the data! All the published results agree that the contaminant neutron patient dose in relation to the tumor dose is about ten times larger for megavoltage x-ray therapy than for electron therapy at the same energy. This is as expected from basic physics and from the fact that the same dose is specified for the tumor in electron therapy as in megavoltage x-ray therapy. For electron therapy, it is of the order of 0.1% of the tumor dose for electrons in the 25-30 MeV energy range. I do not consider this significant, whatever reasonable RBE is selected for the neutrons.

3.9. INTEGRAL DOSE

The concept of integral dose was introduced by Mayneord, when therapy of deep seated tumors was carried out with supervoltage x rays in the 200 kVp energy range. It tried to relate the tumor dose to the total energy deposited in the normal tissue, and ideally the smallest integral dose for a given tumor dose was deemed best.

Unfortunately, this has absolutely no relevance in electron therapy, in my opinion, for the following reason: any non-centrally located tumor is best approached by one single electron beam, which will lead to the smallest ratio of integral dose to normal tissue tumor dose. Any additional electron beam, not being ideal, will lead to a worse ratio of integral dose to tumor dose, with the conclusion that the best treat-

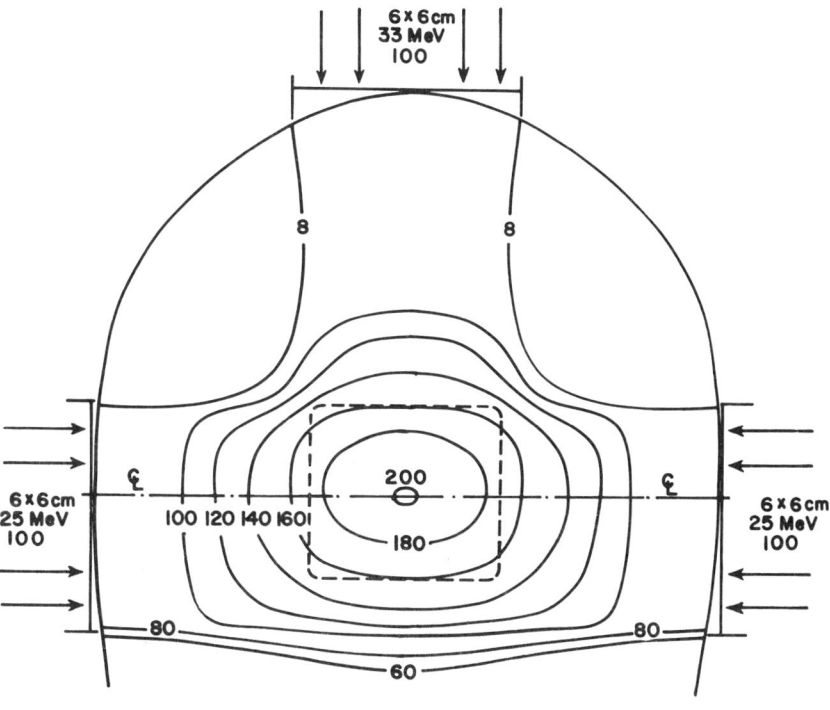

FIG. 3.17. Treatment plan for irradiation of pituitary adenoma. (From Ovadia[3]).

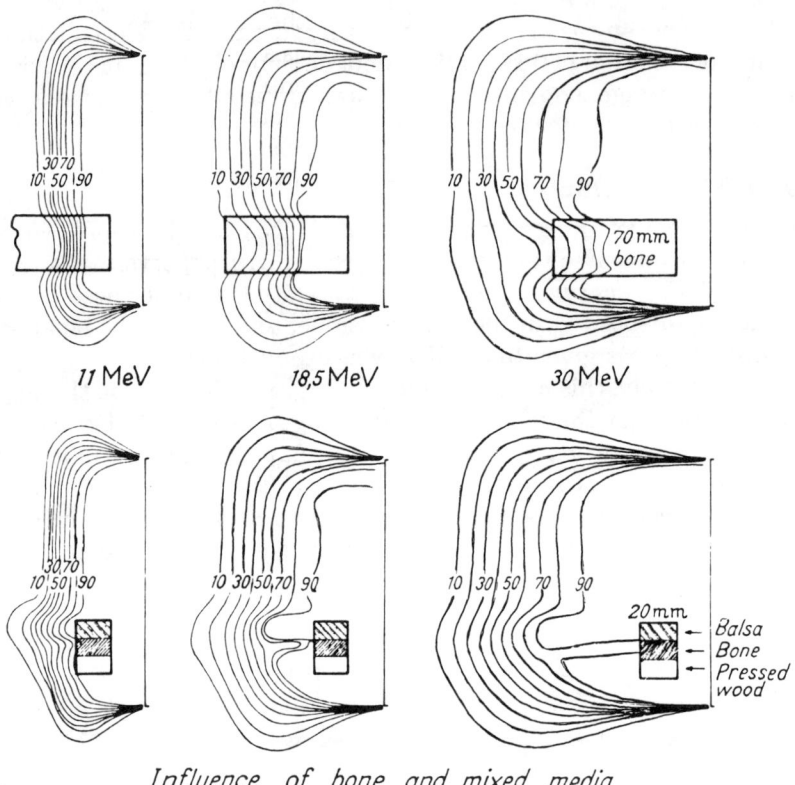

FIG. 3.18. Influence of large volumes of dense bone (upper curves) and a more complex mixture of materials, balsa, dense bone and pressed wood, (lower curves). (From Netteland[1]).

ment occurs with one single field or two parallel opposed fields, which clearly is silly. The problem, as I see it, is that of delivering the prescribed tumor dose, and keeping the dose to specific organs below a certain tolerance. It is not clear to me which is better clinically, a smaller volume of normal tissue irradiated to a higher dose or a larger volume treated to a correspondingly smaller dose, for the same integral dose, and of course making sure that no critical organ, such as the spine or kidney, is irradiated beyond its tolerance. But this is the area of experimental clinical radiation therapy, which is outside the scope of this symposium, and way beyond may competence.

3.10 REFERENCES

[1]*Proceedings of the Symposium on High-Energy Electrons, Montreux, Switzerland, September 7–11, 1964,* edited by A. Zuppinger and G. Poretti (Springer, New York, 1965).

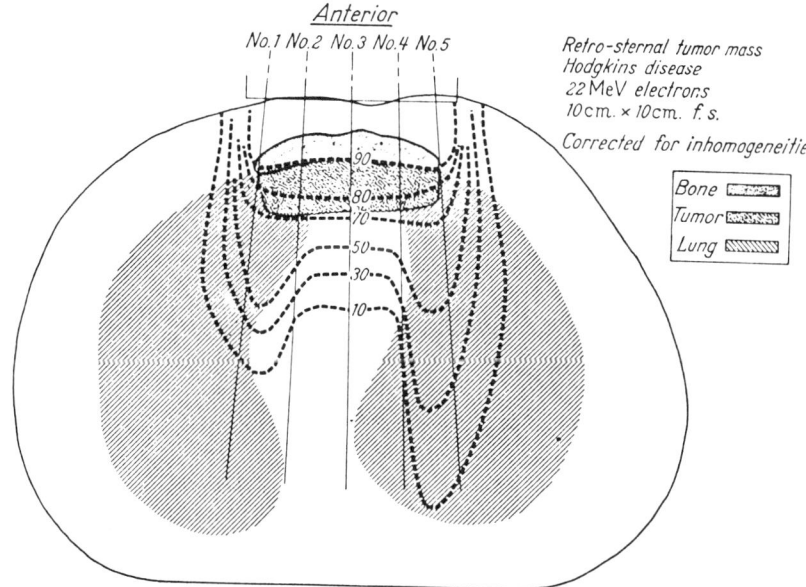

FIG. 3.19. Calculation of isodose distribution with correction for bone and lung inhomogeneity. (From Langhlin[1]).

[2]"Electron Beam Therapy: Proceedings of the Second Annual San Francisco Cancer Symposium, 1966," in *Frontiers of Radiation Therapy and Oncology*, 2, edited by J. Vaeth (S. Karger, Basel, 1969).

[3]*High-Energy Radiation Therapy Dosimetry*, edited by J.S. Laughlin, Ann. N.Y. Acad. Sci. **161**, 1–370 (1969).

[4]*Clinical Applications of the Electron Beam*, edited N. duV. Tapley. (Wiley, New York, 1976).

[5]"High Energy Photons and Electrons: Clinical Applications in Cancer Management," *Proceedings of the International Symposium on the Clinical Usefulness of High-Energy Photons and Electrons (6–45 MeV) in Cancer Management, Philadelphia, May 22–24, 1975*, edited by S. Kramer, N. Suntharalingam, and G. Zinninger (Wiley, New York, 1976).

[6]*Proceedings of the Symposium on High Energy Electrons, Madrid, Spain, 1966*, edited by Gil y Gil and Gil Gayarre (General Directorate of Health, Madrid, 1970).

[7]Florence C.H. Chu, "Electron-Beam Therapy of the Breast,"Radiol. **89** (2), 216–23 (1967).

[8]Florence C.H. Chu, "Experience with Electron Irradiation of Breast Cancer," in *Symposium on High-Energy Electrons*, edited by A. Zuppinger and G. Poretti (Springer, New York, 1963), p. 343ff.

4. Algorithms for computerized treatment planning

Edward S. Sternick, Ph.D.*

Dartmouth-Hitchcock Medical Center, Hanover, New Hampshire

4.1 INTRODUCTION

The importance of computer technology to radiation therapy has been increasingly appreciated since the mid-1950s when it was first proposed that isodose curves be calculated electronically rather than by hand. To date, the preponderance of computations has been directed to either brachytherapy or external photon beams. Although machines capable of producing clinical electron beams have been available for many years, until recently there were few centers in the world with such equipment and most treatments involved only single portals. Hence, there was little incentive to computerize the process. It is interesting to note that, as late as 1974, there was but a single paper dealing with electron beam treatment planning submitted to the Fifth International Conference on the Use of Computers in Radiation Therapy.[1]

Two major events may reverse the apparent lack of concern by physicists in computerizing the treatment planning process for electron beams in the next few years. The first of these involves availability of equipment. An increasing number of radiotherapy departments contemplating the addition of a high energy treatment machine now demand that electrons with a variety of energies be included. Secondly, the complexity of treatment plans currently in use or proposed is growing markedly. Multi-modality treatment (various combinations of photons and electrons), pendulum therapy, and multiple electron fields of several different energies, all involve calculations that when done by hand are tedious and incomplete. If we add the requirement that some type of correction must be done to account for the presence of inhomogeneities such as lung, bone, and air cavities, hand calculations begin to loom as a formidable barrier against the optimal use of the therapy equipment. Thus, it is logical that the computer begin to play a more important role in electron beam treatment planning, just as it has revolutionized the process for photon beams.

It is the intent of this paper to describe a number of procedures, or algorithms, which have been proposed to solve the particular problem of describing an electron beam dose distribution within a patient. The approaches taken by various investigators can be broadly categorized into three groups: 1) empirical, in which large amounts of data taking

---

*Presently at Tufts New England Medical Center, Boston, Massachusetts.

may be required, but which result in a machine-specific formulation having great accuracy; 2) semiempirical, in which data acquisition requirements are reduced by the use of formulae which operate on a minimal number of measurement points, but for which accuracy may be sacrificed; and 3) analytical, in which the dose distributions are modeled by a mathematical expression or expressions derived from basic physical principles.

4.2. EMPIRICAL APPROACHES
4.2.1. Absorption equivalent thickness (AET) method [2-5]

The AET method is based on experimental measurements which indicate that materials of different density in the body have a significant effect on the electron dose distribution. In tissues such as lung with a low density relative to water, reduced attenuation results in an increase of absorbed dose beyond the lung. Although the dose within the lung is also elevated, decreased scattering partially offsets the dose rise. When dealing with high density structures such as bone, it is noted that the increased attenuation results in a displacement of the isodose lines toward the body surface beyond the bone, and a consequent reduction of dose at a specified depth.

A general correction factor called the Absorption Equivalent Thickness (AET) has been described which accounts for the above-mentioned shifts in isodose contours. Although it is recognized that the AET is not constant but is a function of the depth, extent, and characteristics of the inhomogeneity, it was originally suggested that in practice an average correction factor could be used with acceptable accuracy:

$$AET_{(lung)} = 1.3\rho, \tag{4.1}$$

where ρ is the assumed lung density. For example: Assuming a lung density of 0.38g/cm^3, the average correction factor would be:
$$AET_{(lung)} = 1.3(0.38 \text{g/cm}^2) = 0.5. \tag{4.2}$$

When lung density measurements are available for an individual patient, these are incorporated in the determination of AET since it has been shown that the density of lung tissue may be widely varying in the range from 0.2 to 0.9g/cm^3.

An adjustment for bone can also be made. Thus, an AET factor of 1.6 is numerically equal to an actual bone density of 1.85g/cm^3 corrected to the relative number of electrons/cm³ in soft tissue, i.e.,

$$AET_{(bone)} = \frac{1.85 \text{g/cm}^3}{1.0 \text{g/cm}^3} \cdot \frac{(\text{electrons/g in bone})}{(\text{electrons/g in water})} \tag{4.3}$$

$$= \frac{1.85 \text{g/cm}^3}{1.0 \text{g/cm}^3} \cdot \frac{3.0 \times 10^{23} \text{ electrons/g}}{3.36 \times 10^{23} \text{ electrons/g}} \tag{4.4}$$

$$= 1.6$$

The appropriate correction may be made as follows:

1. An isodose plot of selected energy and field size based on unit density measurements is used as a template.
2. On it are projected the central and other diverging rays from the virtual sources at selected intervals across the field.
3. Depth-dose profiles are calculated for each ray with the intersection of isodose lines and contours noted as a function of distance along the ray.
4. An AET is assigned to inhomogeneous tissue structures, and corrections are made to the dose profile for each ray. Beyond the distal boundary of the inhomogeneity, the corrected dose plot will be either above (lung) or below (bone) the uncorrected plot.
5. A corrected isodose distribution is generated by joining equal dose levels on the rays.

4.2.2. Absorption coefficient method[4]

As mentioned in the previous section, AET is not constant for a given energy, but changes with the depth of the inhomogeneity. In order to avoid this variation, the absorption coefficient method makes use of the fact that electron beam depth-dose data show an empirical relationship in any medium, which can be expressed as the simple exponential

$$I(x) = 110 - 10 \exp(\mu x), \tag{4.5}$$

where

$I(x)$ = relative ionization in percent,
x = distance beyond the depth at which $I(x)$ equals 100%,
μ = empirical determined absorption coefficient.

The expression can be further developed to yield

$$I(x) = 110 - 10 \exp(\mu_t x_t + \mu_b x_b + \mu_l x_l), \tag{4.6}$$

where μ is the absorption coefficient respectively for tissue (t), bone (b), and lung (l), and x is the corresponding thickness of the indicated material. The absorption coefficients are dependent on field size, the nature of the material traversed, and electron energy, but not on the depth at which the inhomogeneity is found.

4.2.3. CET method[5-7]

By appropriate measurements either in a suitable phantom or *in vivo* it is possible to define a Coefficient of Equivalent Thickness (CET) which can be used to determine the dose distribution in lung tissue. The CET varies with energy, depth into the inhomogeneity, and density, decreasing with increasing thickness of lung tissue. The correction factor reduces a given thickness of lung tissue to an equivalent absorption thickness of unit density material. To calculate the dose at a given depth in lung, parameters required are the appropriate CET value, chest-wall thickness, and total depth to the point. From this data, corrected doses in lung can be calculated for both central axis and off-axis points except at the periphery of the field.

The CET is defined as (Fig. 4.1)
$$\text{CET} = x_1/x_2 \quad \text{at } y_{2(x_2)} = y_{1(x_1)}. \tag{4.7}$$

The subscripts "1" and "2" respectively refer to the standard water depth-dose curve and the depth-dose curve measured in a water-cork phantom, x is the depth in the phantom measured from the interface, and y is the percentage depth-dose curve at which x_1 and x_2 are measured.

The decreasing part of the depth dose curve in water can be described by Eq. (4.5) or in slightly modified form as

$$y_{1(x_1)} = 110 - 10 \exp[\mu(x+c)], \tag{4.8}$$

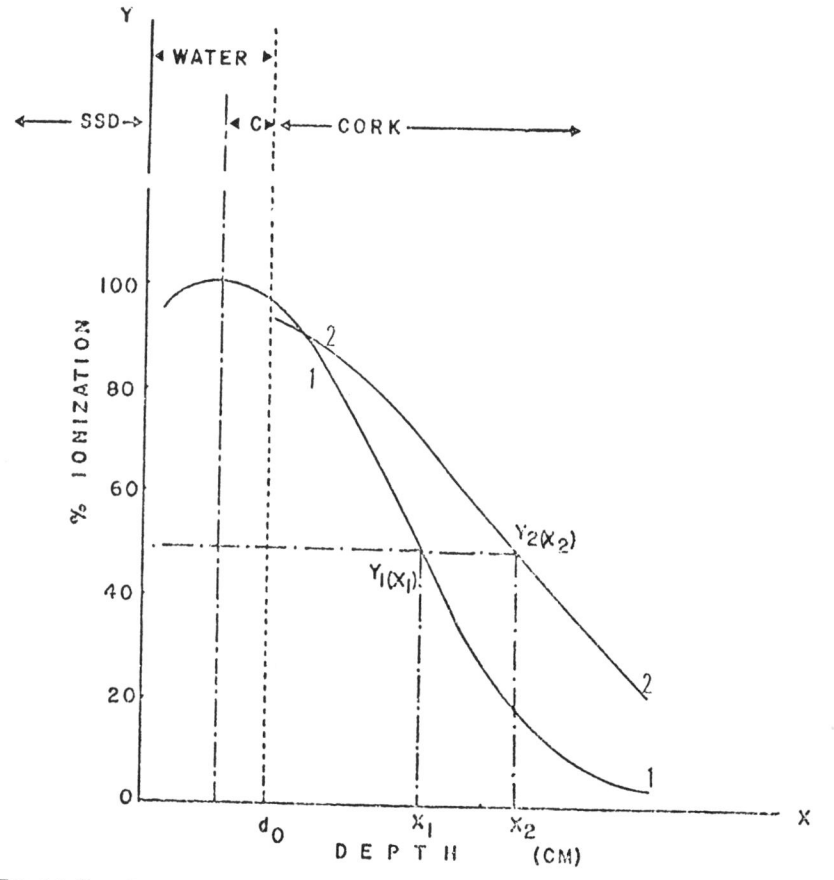

FIG. 4.1. Coordinates and parameters used for the calculation of CET and MAC methods.

where c is the difference between the interface and the depth of the 100% point on the water curve; c may be either positive or negative.

If the CET is introduced into Eq. (4.8), we have

$$y_{2(x_2)} = 110 - 10\exp\mu(x\cdot\text{CET} + c). \tag{4.9}$$

For divergent rays an inverse square correction factor is introduced to compensate for the reduction of radiation intensity with increasing distance from the radiation source given by

$$S = \left(\frac{\text{SSD} + d_0 + \text{CET}\cdot x_2}{\text{SSD} + d_0 + x_2}\right)^2. \tag{4.10}$$

Where

$$y_{2(x_2)} = y_{1(x_1)}, \tag{4.11}$$

it follows that

$$y_{2(x_2)} = [110 - 10\exp\mu(x_2\cdot\text{CET} + c)]\left(\frac{\text{SSD} + d_0 + \text{CET}\cdot x_2}{\text{SSD} + d_0 + x_2}\right)^2. \tag{4.12}$$

Although Eq. (4.12) has been derived for doses along the central axis, off-axis points can also be calculated. Measurements have indicated that CET values are relatively independent of the parameters d_0, c, and μ when the chest wall is on the order of 1–2 cm thick. As in the AET method described, the beam can be divided into separate rays, a new attenuation coefficient μ and values for x and d_0 measured, and the corrected doses calculated along each ray.

4.2.4. MAC method[8]

The Modified Absorption Coefficient (MAC) method further extends the techniques described above to make additional corrections for the polarization effect, the deviation of the central axis percentage depth dose from the relationship of Eq. (4.5) for energies above 20 MeV and the variation in densities of internal organs for a specific patient.

For a given percentage depth dose measured in a water-cork phantom at a distance x_2 cm from the interface, MAC is defined as (Fig. 4.1)

$$\text{MAC} = x_1\rho_1/x_2\rho_2, \tag{4.13}$$

where

$$y_2(x_2) = y_1(x_1)S \tag{4.14}$$

and

$$S = \left(\frac{\text{SSD} + d_0 + \text{MAC}\cdot x_2\rho_2/\rho_1}{\text{SSD} + d_0 + x_2}\right)^2. \tag{4.15}$$

The water and cork densities are designated by ρ_1 and ρ_2 respectively and S is the inverse square correction factor.

Although percentage depth-dose curves for energies ⩽20 MeV follow the relationship of Eq. (4.5), data obtained for higher energies deviate significantly from a straight line at values below the 50% percentage depth-dose level (Fig. 4.2). The corrected relationship is expressed by

$$I(x) = 110 - 10 \exp\{\mu[(x-d_m)\rho - \lambda(x-d_m)^3\rho^3]\}, \quad (4.16)$$

where ρ is the density of the medium ($\rho = 1$ for water) and λ is the depth-dose correction coefficient. Regardless of field size, λ is constant for a given material and energy. For energy ⩽20 MeV, λ approaches zero.

The determination of μ usually requires measurements at various depths for each energy and field size to be used. The number of such measurements can be reduced by further development of Eq. (4.5),

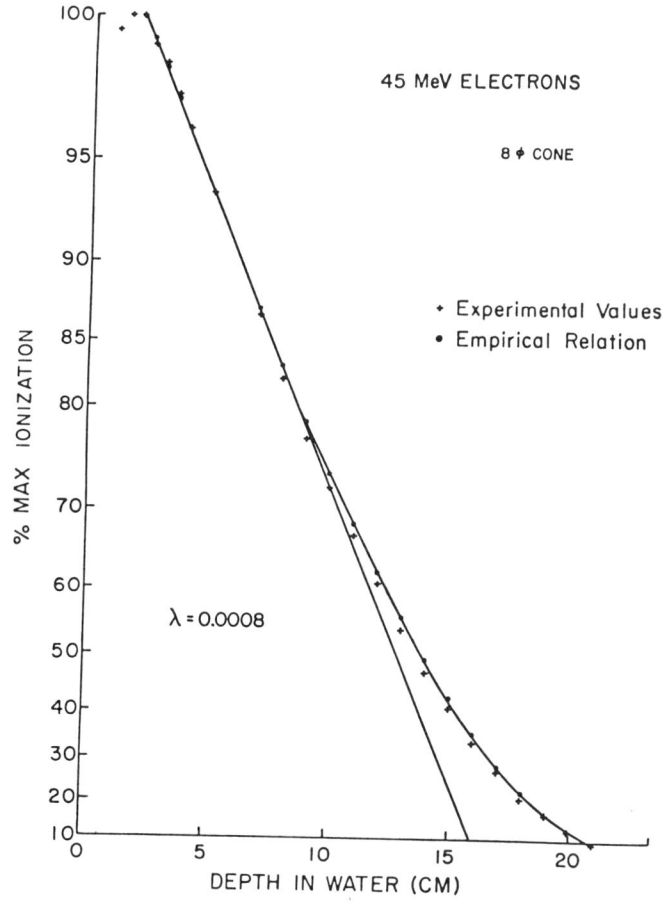

FIG. 4.2. Percentage ionization curve for 45 MeV electrons showing deviation from a straight line.

which can also be written as

$$I(x) = 110 - 10 \exp[\mu(x - d_m)], \tag{4.17}$$

where x is the depth in the phantom and d_m is the maximum build-up depth.

To simplify the calculation of μ, let

$$\mu = 0.693/(d_{90} - d_m), \tag{4.18}$$

where d_{90} is the depth where the depth dose is 90%. The depth in water can then be expressed in units of $d_{90} - d_m$ and an absorption number (n) defined, where

$$n = (x - d_m)/(d_{90} - d_m). \tag{4.19}$$

Substituting Eqs. (4.18) and (4.19) into Eq. (4.17) gives the following results

for $n = 1$, $x = d_{90}$, $I(x) = 90\%$ depth dose,
for $n = 2$, $x = d_{70}$, $I(x) = 70\%$ depth dose,
for $n = 3$, $x = d_{30}$, $I(x) = 30\%$ depth dose.

Then, by using the relationships derived above we can regenerate Eq. (4.16) into a slightly different form:

$$I(x) = 110 - 10 \exp\{0.693n\rho[1 - \lambda(x - d_m)^2\rho^2]\}. \tag{4.20}$$

This implies that only a few measurement points will be required to determine the depth of the 90% level and consequently the absorption number for various energies and field sizes.

The percentage depth dose at a given point in water is obtained as

$$y_1(x_1) = (110 - 10 \exp\{0.693n\rho_1[1 - \lambda(x_1 - c)^2\rho_1^2]\})P, \tag{4.21}$$

where P is the correction for the polarization effect and c is the distance between the 100% depth dose point and the surface–interface distance.

Then, MAC and Eq. (4.16) may be used to express $y_2(x_2)$ as

$$y_2(x_2) = \{110 - 10 \exp[0.693\text{ED}/(d_{90} - d_m)]\}SP, \tag{4.22}$$

where

$$\text{ED} = \text{effective depth} = (\text{MAC}\rho_2 x_2 - c\rho_1) - \lambda(\text{MAC}\rho_2 x_2 - c\rho_1)^3. \tag{4.23}$$

4.2.5. Equivalent thickness methods and edge effects[9-11]

The above techniques have been developed assuming that the inhomogeneities encountered are large and uniform slabs. When the inhomogeneities in the path of the electron beam are small, accurate modification of the dose distribution is not easily accomplished. The following presents an empirical approach to this problem which should give correct dose values to within 10–15%.

Consider Fig. 4.3, which shows a measured dose distribution behind a lead strip. There is an overdosage of approximately 20% which is distal to the edge and a corresponding dose reduction medial to the edge. The situation is analogous to the dose distribution behind a small bone.

The angle β defines the limit of the area influenced by scattered electrons from the edge and is dependent on the mean electron energy $\bar{E}$:

$$\bar{E} = E_0(1 - z/R_p), \tag{4.24}$$

where
E_0 = initial electron energy,
R_p = practical range,
z = depth in material.

The maximum change, P_{max}, of absorbed dose due to an edge between different kinds of material is

$$P_{max} = \frac{D_{max} - D_0}{D_0}, \tag{4.25}$$

where D_{max} is the absorbed dose at the highest increase or depression as

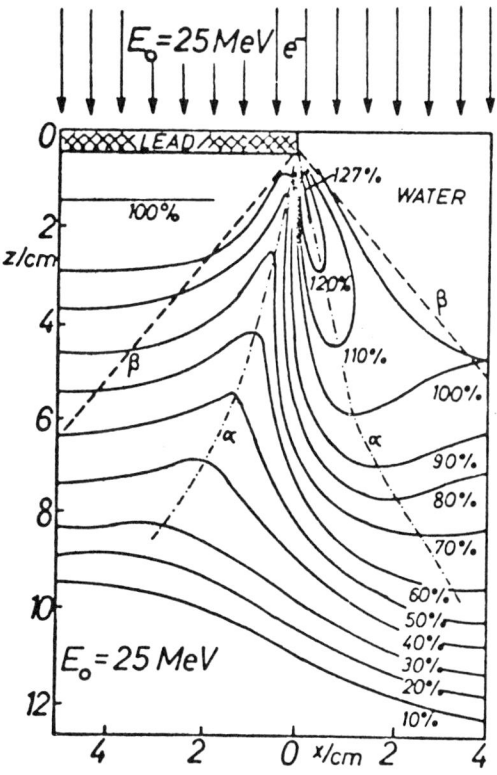

FIG. 4.3. Measured dose distribution behind a lead strip in water. (From Pohlit.[11])

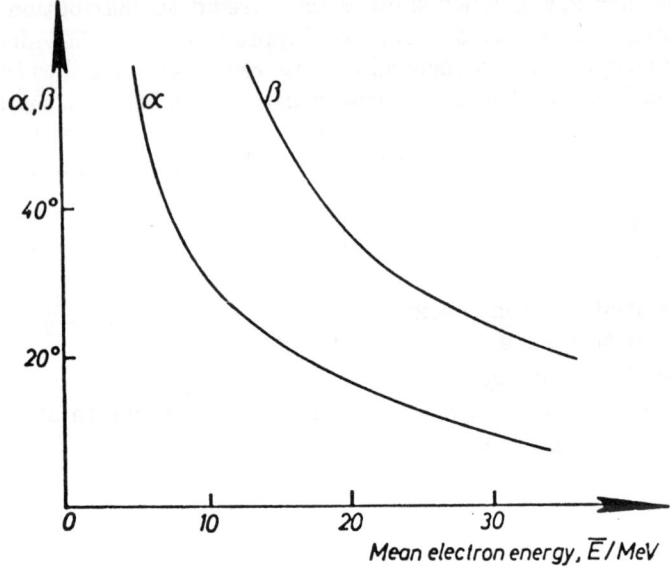

FIG. 4.4. Energy dependence of the angles α and β. (From Pohlit.[11])

a result of scatter and D_0 is the absorbed dose in the absence of such scatter at the same point.

The method for correcting the isodose distribution then proceeds as follows:

1. First, correct the dose distribution for large slab-type inhomogeneities.

2. If edges are present in the distribution, find the mean electron energy at the edge and determine the angles α and β (Fig. 4.4).

3. Finally, determine the relative increase and decrease of absorbed dose along angle α at different distances from the edge (Fig. 4.5).

4.2.6. Calculations for pendulum therapy with simple depth-dose formulae[12,13]

Calculations of dose distributions for moving beam electron therapy can be made in a straightforward manner by using measured values of depth doses and calculations of a correction factor for the lateral fall-off of dose away from the central axis.

In a homogeneous tissue irradiated by an electron beam of a given energy, field size the source-surface distance, the dose distribution in the principal plane is given by

$$D(x,y) = D(x_m, 0) \cdot T(x) \cdot Q(x,y) \cdot 10^{-4}, \tag{4.26}$$

where

$D(x_m,0)$ = maximum dose on the central axis,
$T(x)$ = percentage depth dose,
$Q(x,y)$ = off-center ratio.

If the field is moving, the dose D_{kn} at the point within the patient is:

$$D_{kn} = \int_0^{t_1} D_{kn}\, dt = \frac{1}{w}\int_{\alpha_1}^{\alpha_2} D_{kn}\, d\alpha, \qquad (4.27)$$

where

t_1 = irradiation time,
w = angular velocity,
α_1 and α_2 = limits of the rotation angle.

Equation (4.27) can be approximated by the following:

$$D_{kn} = \frac{\Delta\alpha}{w}\sum_i D_{ikn} \qquad (4.28)$$

$$= \frac{\Delta\alpha}{w}\sum_i D_i(x_m,0)\cdot T_i(x)\cdot Q(x,y)\cdot 10^{-4}, \qquad (4.29)$$

where

$$\Delta\alpha = \frac{\alpha_2-\alpha_1}{n}.$$

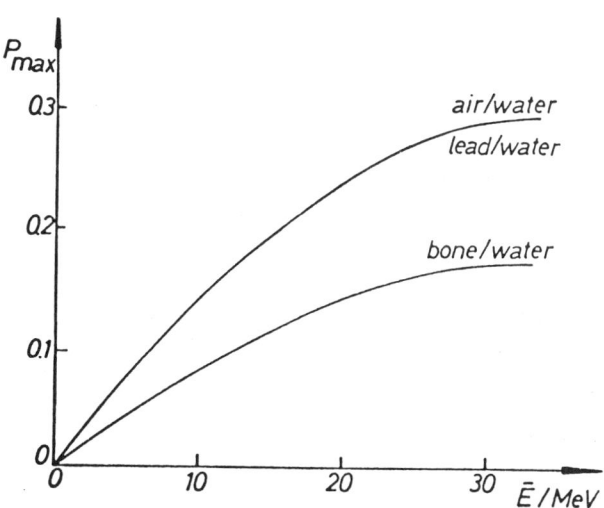

FIG. 4.5. Maximum changes in absorbed dose behind edges. (From Pohlit.[11])

4.3. SEMI-EMPIRICAL APPROACHES

4.3.1. Approximation by the Age diffusion method[14-17]

To reduce the amount of data required for accurate characterization of the electron beam, a method based on a general age diffusion equation has been proposed and successfully implemented for both a betatron and linear accelerator.

The development of this equation leads to

$$D(x,y,z,\tau) = \frac{D_0}{2}\left[\text{erf}\left(\frac{x_0(z)-x}{2(k\tau)^{1/2}}\right) + \text{erf}\left(\frac{x_0+x}{2(k\tau)^{1/2}}\right)\right]$$

$$\cdot\left[\text{erf}\left(\frac{y_0(z)-y}{2(k\tau)^{1/2}}\right) + \text{erf}\left(\frac{y_0+y}{2(k\tau)^{1/2}}\right)\right]$$

$$\cdot\cos[G_1(z/R_p)^2 + G_2(z/R_p) + G_3]\cdot[F/(F+z)]^2$$

$$\cdot\exp[-(2\pi/3R_p)\cdot(k\tau)^{1/2}]^2, \qquad (4.30)$$

where

$D(x,y,z,\tau)$ = dose at depth z,
$(k\tau)^{1/2} = (C_z/R_p + P)^N$,
F = source-skin distance,
$x_0 = \frac{1}{2}(\text{width} + K_e)\cdot[(F+z)/F]$,
$y_0 = \frac{1}{2}(\text{length} + K_e)\cdot[(F+z)/F]$,
R_p = practical range,
$\left.\begin{array}{l}G_1 \\ G_2 \\ G_3 \\ C \\ N \\ P\end{array}\right\}$ constant for a given energy,
K_e = field size correction for edge effects,
$k\tau$ = age diffusion parameter.

The variation of the percentage depth dose with field size is predicted by the error function factors which allow dose predictions to be made for any point in a plane, for rectangular beams, either on or off the central plane.

Beam shapes are determined by the shape parameters C, N, and P, and they are strongly interrelated. P characterizes the beam diffuseness at the surface; C characterizes the increase in diffuseness with depth; and N is an overall shape-changing parameter. It is also necessary to make small modifications to the width and length of rectangular beams with the constant, K_e, in order to get a good fit to the experimental data.

The parameters G_1, G_2, and G_3 can be calculated from:
1) the measured central axis depth-dose data, z and $D(z)$,
2) the practical range, R_p,

3) the field width and length, x and y,
4) the source-surface distance, F,
5) the constants, C, N, P.

In practical use, the parameters are determined at each energy for one field size and then are used to predict the depth doses and isodoses for all field sizes. The method also allows modification of the generated dose distributions to account for the presence of inhomogeneities.

4.3.2. Pencil beam approximation[18,19]

The dose distribution in a broad, parallel electron beam entering a uniform phantom can be resolved into a series of narrow or pencil beams. It is thus possible to build up a broad-beam dose distribution from a knowledge of the axial and radial dose distributions in the pencil beam and to construct isodose distributions in both homogeneous and heterogeneous phantoms.

The procedure is the following:
1. Measure the narrow-beam dose distributions in planes at a right angle to the beam axis.
2. Express the results in the form of a dose matrix in each plane in which measurements are made.
3. Calculate the dose distribution in these planes for broad beams by summation of the narrow-beam matrices at appropriate spacings.
4. From the dose distribution calculated in (3), construct isodose curves for the broad beam.

Narrow beams of small diameter (~2.5 mm) are produced by a special collimator and the dose distributions measured with film sandwiched between sheets of Lucite or aluminum placed normal to the longitudinal axis of the beam.

4.3.3. The formula of Czaikowsky[20,21]

The dose at a depth z in tissue can be described by a formula originally developed by Czaikowsky:

$$D(z) = D_p \exp[-\alpha(z/R_p)]^\beta + D_s d_s(z/z_m), \qquad (4.31)$$

where
D_p = dose contribution of primary electrons at the dose maximum,
D_s = dose contribution of secondary electrons at the dose maximum,
z_m = depth of dose maximum,
R_p = practical range,
$d_s(z/z_m)$ = function of depth.

The coefficients α and β and the practical range, R_p, are dependent on the primary energy of the electrons and the field size and type of scattering foil used.

Dose distributions off the central axis can be calculated by using a small number of numerically or analytically described transverse distributions for a small number of field sizes, beam energies, and depths, interpolating for intermediate values.

4.3.4. Approximation by the difference method[22,23]

The principle of representation used by the difference method consists of a mathematical description of the differences between actual dose distributions for a given machine and "ideal" isodose distributions which for a single field isodose line can be defined on the basis of three primary relationships: 1) Region I (near the center) is described by the classical relation of the absorbed dose along the central axis of the beam; 2) Region II (edges) is based on a relation which reveals the change in dose as a function of the distance to the central axis at a constant depth; and 3) Region III (transition between center and edges) indicates the transition between Regions I and II.

For large fields the depth z on the central axis for a dose D at an energy E is given by:

$$z = \left(0.52 - 0.18D - k_1 \left(\frac{0.01E}{1-D}\right)^2\right) E - 0.2 - k_2 D E^2, \tag{4.32}$$

where
$k_1 = 0.010$ for D between 0.1 and 0.8
$k_1 = 0.0053$ for $D = 0.9$,
$k_2 = 0$ for wide fields

and

$k_2 \approx 0.001$ for a 7 cm $\times$ 7 cm field.

The complete isodose curve can be described by

$$x = \frac{L}{2}\left(\frac{\text{SSD}+z}{\text{SSD}}\right) + k_3(k_4 - D)\left(\frac{f+z}{f+E/5}\right) - \frac{0.04}{(1.1 - z/z_{\max})^2}, \tag{4.33}$$

where x is the distance to the axis of the isodose curve D; this distance varies with the depth z, the field size, L, and the source to phantom distance, SSD. k_3 and k_4 are constants describing the penumbra, and

$$\Delta x = k_3(k_4 - D), \tag{4.34}$$

where Δx is the distance between the isodose curves and the theoretical field limit,

$\left.\begin{array}{l}k_3 = 3.49 \\ k_4 = 0.57\end{array}\right\}$ without diaphragm,

$\left.\begin{array}{l}k_3 = 2.34 \\ k_4 = 0.53\end{array}\right\}$ with diaphragm.

4.4 ANALYTICAL APPROACHES[24-29]

Analytical approaches rely on modeling of the electron beam and must be regarded in an evolutionary stage. The methods have been applied to complex discontinuities as well as to single field irradiation of homogeneous phantoms, but routine clinical use of these algorithms is not widespread at this time.

4.4.1 Central axis calculation

A simple analytical formula for the central axis percentage depth dose for electrons in the 5–30 MeV range is the following:

$$D(x) = 100 \left(\frac{R - x}{R - x_m} \right)^a \left[1 + a \left(\frac{x - x_m}{R - x_m} \right) \right] \tag{4.35}$$

where
  $D(x)$ = central axis percentage depth dose,
  R = the range of electrons of initial energy E_0,
  x = penetration depth,
  $a = \mu_0 R / E_0$,
  x_m = depth of dose maximum.

The formula can be generalized to include situations involving different density layers and varying field sizes. It can be expanded to allow for transverse distributions, and the calculation of full isodose curves.

4.4.2. Pencil beam method

A theoretical approach to the pencil beam method discussed previously has also been developed. If $D_p(\phi)$ is the dose at any point P at a distance d off the central axis of the beam, then $D_p(\phi)$ can be expressed as an integral between α_1 and α_2 (Fig. 4.6):

$$D_p(\phi) = \frac{1}{(2\pi)^{1/2}} \int_{\alpha_2}^{\alpha_1} \frac{1}{\sigma} \exp\left(-\frac{\alpha^2}{2\sigma^2} \right) d\alpha, \tag{4.36}$$

where

$$\alpha_1 = \tan^{-1}\left(\frac{d-l}{x}\right) = \tan^{-1}\left(\tan\phi - \frac{l}{x}\right),$$

$$\alpha_2 = \tan^{-1}\left(\frac{d+l}{x}\right) = \tan^{-1}\left(\tan\phi - \frac{l}{x}\right),$$

  l = half-field width at surface,
  σ = standard deviation defined as the square root of the average of the squares of the angular deviations from the mean,

and

$$\sigma^2 = \frac{4\pi N^2 Z^2 e^4 x}{P_x^2 V_x^2} \cdot \ln\left(\frac{137 P_x}{Z^{1/2} mc}\right), \tag{4.37}$$

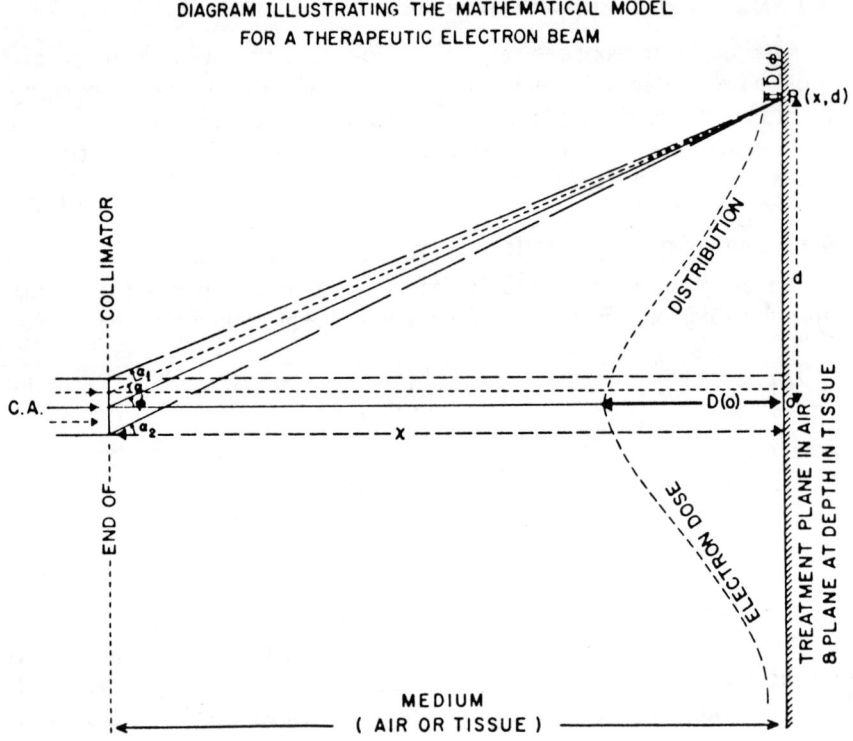

FIG. 4.6. Mathematical model for a pencil beams. (From Osman[27].).

where
 N = number of scattering centers per cm^3,
 Z = effective atomic number of material traversed by incident electrons,
 P_x = average momentum of electrons at a depth x,
 V_x = average velocity of electrons at a depth x,
 m = rest mass of the electron,
 c = speed of light.

P_x and V_x can be obtained from the energy-range relations:
$$\rho R_p = 0.530 E_0 - 0.16 \tag{4.38}$$

or
$$\rho R_p = 0.521 E_0 - 0.376, \tag{4.39}$$

where
 R_p = practical range,
 ρ = density of the medium,
 E_0 = energy at the surface.

Then
$$E_x = E_0(1 - x/R_p) + 0.511 = (c^2 P_x^2 + m^2 c^4)^{1/2}, \quad (4.40)$$

where

E_x = average energy of electrons at depth x

and
$$V_x = c^2 P_x / E_x. \quad (4.41)$$

4.5. CONCLUSIONS

A number of published algorithms are available for implementation in computerized electron-beam treatment planning systems. Because our experience to date is limited, there is not a direct answer to the question, "Which one is best?" Potential users of such algorithms must be prepared to undertake a comprehensive analysis of their own local needs to ascertain the suitability of one or another of these approaches. Certainly, extensive direct measurements are mandatory on the user's own linear accelerator or betatron to verify the goodness-of-fit of the algorithm selected. It is not unexpected, perhaps, that the accuracy of calculation, particularly where inhomogeneities are involved, is strongly dependent on the algorithm one chooses to adopt for the specific electron accelerator used.[30]

4.6. REFERENCES

[1] *Computer Applications in Radiation Oncology*, edited by E.S. Sternick (Univ. Press of New England, Hanover, N.H., 1976).

[2] J.S. Laughlin, "High Energy Electron Treatment Planning for Inhomogeneities," Brit. J. Radiol. **38**, 143–47 (1965).

[3] J.S. Laughlin, A. Lundy, R. Phillips, F. Chu, and A. Sattar, "Electron-Beam Treatment Planning in Inhomogeneous Tissue," Radiology **85**, 524–30 (1965).

[4a] A. Dahler, A.S. Baker, and J.S. Laughlin, "Comprehensive Electron-Beam Treatment Planning," Ann. N.Y. Acad. Sci. **161**, 198–213 (1969).

[4b] J.G. Holt, R. Mohan, D. Perry, L. Simpson, and J.S. Laughlin, "Memorial Computerized Electron-Beam AET Treatment Planning System," submitted to the Sixth International Conference on the Use of Computers in Radiation Therapy, Gottingen, Germany, September, 1977.

[5] M.L.M. Boone, J.H. Jardine, A.E. Wright, and N.D. Tapley, "High Energy Electron Dose Perturbations in Regions of Tissue Heterogeneity Part I: In Vivo Dosimetry," Radiology **88**, 1136–45 (1967).

[6] P.R. Almond, A.E. Wright, and M.L.M. Boone, "High Energy Electron Dose Perturbations in Regions of Tissue Heterogeneity Part II: Physical Models of Tissue Heterogeneities," Radiology **88**, 1146–53 (1967).

[7] M.L.M. Boone, P.R. Almond, and A.E. Wright, "High Energy Electron Dose Perturbations in Regions of Tissue Heterogeneity," Ann. N.Y. Acad. Sci. **161**, 214–32 (1969).

[8] F. Bagne, "Electron Beam Treatment Planning System," Med. Phys. **3**, 31–38 (1976).

[9] W. Pohlit, "Calculated and Measured Dose Distributions in Inhomogeneous Materials and in Patients," Ann. N.Y. Acad. Sci. **161**, 189–97 (1969).

[10] W. Pohlit, "Treatment Planning for Electron Beams in Inhomogeneous Media," in *Radiation Dosimetry*, AAPM 1976 Summer School, pp. 81–86.

[11] W. Pohlit and K.H. Mangold, "Electron-Beam Dose Distribution in Inhomogeneous Media," in *High Energy Photons and Electrons*, edited by S. Kramer, N. Suntharalingam, and S. Zinninger (Wiley, New York, 1976), pp. 245–54.

[12] G.R. Benedetti, H. Dobry, and L. Taumann, "Computer Programme for Determination of Isodose Curves for Electron Energies from 5 to 42 MeV," Electromedica **2**, 57–60 (1971).

[13] H.K. Leetz, "Calculation of Dose Distributions in 5–42 MeV Betatron Electron Beams for Treatment Planning," Digest of the Fourth International Conference on Medical Physics, Special Issue of Physics in Canada **32**, 23.5 (1976).

[14] K. Kawachi, "Calculation of Electron Dose Distribution for Radiotherapy Treatment Planning," Phys. Med. Biol. **20**, 571–77 (1975).

[15] J.D. Steben, K. Ayyangar, and N. Suntharalingam, Digest of the Fourth International Conference on Medical Physics, Special Issue of Physics in Canada **32**, 28.10 (1976).

[16] J.D. Steben, K. Ayyangar, and N. Suntharalingam, "Betatron Electron Beam Characterization for Dosimetry Calculations," submitted to Phys. Med. Biol.

[17] N.A. Iverson, G.S. Ibbott, R.K. Cacak, and W.R. Hendee, "An Evaluation of the EMI-Rad 8 Electron Beam Simulation Program," presented at the 19th Annual Meeting of the AAPM, Cincinnati, Ohio, August, 1976, Paper No. P2.

[18] J.W. Boag and S.C. Lillicrap, "Dose Distributions in High Energy Electron Beams," in *Digest of the Third International Conference on Medical Physics, Including Medical Engineering*, edited R. Kadefors, R.I. Magnusson, and I. Petersen, Paper 11.1, 1972.

[19] S.C. Lillicrap, P. Wilson, and J.W. Boag, "Dose Distributions in High Energy Electron Beams: Production of Broad Beam Distributions from Narrow Beam Data," Phys. Med. Biol. **20**, 30–38 (1976).

[20] D. Fehrentz, "Calculation of Dose Distributions for Electron Depth Therapy Paying Regard to Inhomogeneities," in *Computer Applications in Radiation Oncology*, edited by E.S. Sternick, (Univ. Press of New England, Hanover, N.H., 1976), pp. 184–89.

[21] D. Fehrentz and B. Kimmig, "Calculation of Dose Distributions for Electrons up to 42 MeV," Digest of the Fourth International Conference on Medical Physics, Special Issue of Physics in Canada **32**, 23.6 (1976).

[22] C. Tronc and M. Levaillant, "Doses Absorbees pour Faisceaux d'Electrons du Sagittaire Mesvres. Systematiques et Leur Representation pour Introduction dans un Calculateur," Brit. Inst. Radiol. Special Report 13, Part 2, 103–04 (1976).

[23] D. Tronc and M. Levaillant, "Absorbed Dose Distributions for Electron Beam from the Sagittaire. Systematic Measurements and Their Representation for Introduction into a Computer," submitted to Strahlentherapie.

[24] J.M. Pacyniak and A. Pagnamenta, "Central Axis Percentage Depth-Dose for High Energy Electrons," Radiation Res. **60**, 342–46 (1974).

[25] F. Borger, L. Simpson, and J. Ovadia, "Electron Beam Treatment Planning Program Using an Analytic Representation of the Electron Beam," Digest of the Fourth International Conference on Medical Physics, Special Issue of Physics in Canada **32**, 23.3 (1976).

[26] G. Osman and R. Johnson, "Theoretical Generation of Isodose Curves and Computerization of Treatment Planning in Electron Therapy," Digest of the Fourth International Conference on Medical Physics, Special Issue of Physics in Canada **32**, 23.4 (1976).

[27] G. Osman, "Dose Distribution of Therapeutic Electron Beams and Automation of Treatment Planning," J. Med. **7**, 143–67 (1976).

[28] M. Goitein, G.T.Y. Chen, J.Y. Ting, R.J. Schneider, and J.M. Sisterson, "Measurements and Calculations of the Influence of Thin Inhomogeneities on Charged Particle Beams," Med. Phys. **5**, 265–73 (1978).

[29] M. Goitein, "A Technique for Calculating the Influence of Thin Inhomogeneities on Charged Particle Beams," Med. Phys. **5**, 258–64 (1978).

[30] J. Ting and E.S. Sternick, "A Comparative Study of Five Tissue Inhomogeneity Correction Algorithms Used in Electron Beam Treatment Planning," Presented at 62nd Meeting of the RSNA, Chicago, 1976, Paper #205.

5. Memorial electron beam AET treatment planning system*

J. Garrett Holt, B.A., Radhe Mohan, Ph.D.,
Richard Caley, Alfonso Buffa, B.S., Ann Reid, B.Sc.,
Larry D. Simpson, Ph.D., and John S. Laughlin, Ph.D.

Memorial Sloan–Kettering Cancer Center, New York, New York

5.1 INTRODUCTION

This is a report on the computerization of the original Memorial Electron Beam Treatment Planning System. The system has now been in use for over 20 years and has been employed in the comprehensive calculation of the dose distribution produced in patients by electrons. These treatment plans are designed on an individual basis, and up to this time the number of comprehensive treatment plans generated is in excess of 5000. The system was originally designed to accommodate completely for curvature and the associated air gap, any inhomogeneity in patient structure relative to unit density, and any beam wedges or bolusing material. This was accomplished through the "ray" system subsequently described in the literature[1-4] together with the absorption equivalent thickness (AET) factor. Depth-dose data, corrected to infinite SSD in water and inhomogeneous tissue, are required to define AET factors. At an arbitrary depth in inhomogeneous tissue, a corresponding depth in water can be found where an equal amount of energy is absorbed per gram. The ratio of the water equivalent depth relative to the inhomogeneous tissue depth is the AET factor.

As an instance of its importance, before its use the incidence of fibrosis or pneumonitis in conventional treatment of chest wall nodes approximated 6% at Memorial Hospital and even more in other institutions. Since the employment of the accurate accommodation of inhomogeneity by the use of the AET factors, such incidence has been virtually eliminated. The original system and its computerization reported here is based directly on experimentally determined data, rather than on the employment of empirical factors. The correction for cavity ionization to dose includes the polarization effect explicitly which has been a feature of this system since the original experimental determination of the validity of the polarization hypothesis for electrons.[5,6]

---

*In the original Symposium, this paper was presented by L. Simpson as part of the Discussion session.

5.2. COMPUTATION MODEL

The model employed in the computer aided calculation of electron beam dose distributions utilizes table look-up and interpolation of measured central axis depth doses and off-center ratios. Inhomogeneity correction is applied by the AET method.

As before, the dependence of AET on depth in tissue and in the inhomogeneity, on field size, density and energy is taken into account. The curvature correction is applied by interpolating between two or more sets of data measured for different air gaps. Distance dependence is accounted for by defining a virtual source and assuming an inverse square law, or optionally, by using the ratio of the doses measured "in air."

To illustrate the formalism of the model, the problem is divided into two parts: correction for curvature and correction for inhomogeneity.

Referring to Fig. 5.1(b), to calculate dose at P along the ray at depth d at a distance x from the central axis, we assume that it is equivalent to the dose at P for a flat surface AQB. The dose at P in terms of central axis depth dose (D_c) and off-center ratio (OCR) is then given by

$$D(d,x,\phi,g) = D_c(d_c,\phi,g_c) * \text{OCR}(d_c,\phi,g_c,x/H) * D_c(d_{max},\phi,g_c)/D_c(d_{max},\phi,0),$$
(5.1)

where d_c and g_c are the projections of d and g on the central axis. Central-axis depth dose D_c is normalized to the maximum for each set. OCR

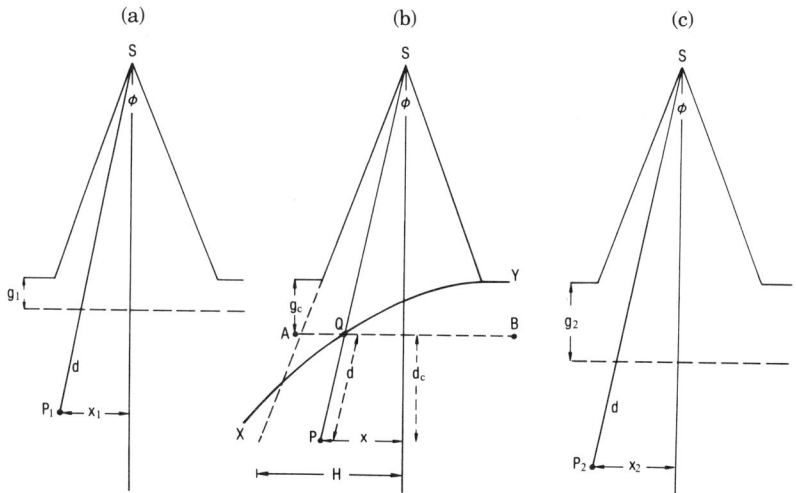

FIG. 5.1. Electron beam dose computation model. In Fig. (b), X, Q, Y is the skin surface and Q is the point of intersection of ray SP and skin. Vertical distance between the skin surface and the cone is g_c. The dose at point P at depth d along the ray and air gap g_c is obtained by interpolating between the dose at point P_1 and air gap g_1 [Fig. (a)] and a point P_2 and air gap g_2 [Fig. (c)].

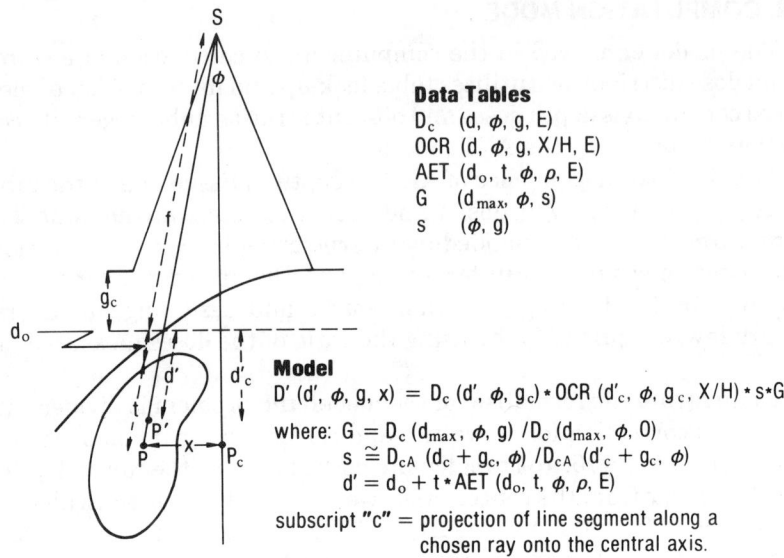

FIG. 5.2. Correction for inhomogeneity is applied by the AET method. Equivalent depth in tissue $d' = d_o + \text{AET} \cdot t$ is calculated, where d_o is the depth of overlying tissue and t is the path length in lung. Point of calculation P is shifted to point P' at depth d' and dose determined at P' by table look-up and interpolation. Since the equivalent depth in tissue does not correspond to the actual location of point of computation, correction for distance dependence of dose is then applied to obtain dose at P.

at a point at depth d is defined as the ratio of dose at that point to the dose at the central axis at the depth d_c, where d_c is the projection of d on the central axis and d is the depth along the ray. OCR is a function of depth, field size, air gap and is expressed in terms of distance x of the point from the central axis as a fraction of half width H at depth d (i.e., x/H).

Values of D_c and OCR in expression (5.1) are obtained by interpolating between the data tables for the two adjacent air gaps g_1 and g_2 as shown in Fig. 5.1(a) and (c). Measured data for air gaps of 0, 3, 6, and 9 cm are available.

To apply correction for inhomogeneity, we define a quantity

$$d' = d_o + t \cdot \text{AET}(d_o, t, \phi, \rho, E), \tag{5.2}$$

where d_o is the depth in tissue up to the inhomogeneity, t is the path length in the inhomogeneity along the ray. Thus $d = d_o + t$. AET (absorption equivalent thickness factor) is used to determine the path length d' that includes the inhomogeneity correction. Dose at P is then obtained by shifting P to P' (see Fig. 5.2). P' is at depth d' along the ray SP from the virtual source. Furthermore, correction for distance dependence has to be applied. If we assume an inverse square law from a virtual source, then this correction takes the form of a multiplicative factor:

$$s = [(f+d')/(f+d)]^2$$
$$= [(ssd+d'_c+g_c)/(ssd+d_c+g_c)]^2, \quad (5.3)$$

where f is the distance along the ray to the skin surface.

Alternatively, the distance dependence can be accounted for by multiplying the dose at the shifted point P' by the ratio of doses in air at P and P' respectively, in which case the correction factor will be

$$s = D_A(d+g,\phi)/D_A(d'+g,\phi)$$
$$= [D_A(d_c+g_c,\phi)*OCR_A(d_c+g_c,\phi,x/H)]$$
$$/[D_A(d'_c \mid g_c,\phi)*OCR_A(d'_c+g_c,\phi,x/H)], \quad (5.4)$$

where the subscript A denotes measurements in air. If the off-center ratios in Eq. (5.4) can be assumed to be approximately equal, the correction factor reduces to

$$s = D_{cA}(d_c+g_c,\phi)/D_{cA}(d'_c+g_c,\phi). \quad (5.5)$$

We have available the measured central-axis data in air for all energies and field sizes used in treatments.

If the dose in air on the central axis varied strictly as inverse of the square of the distance from the virtual source, Eqs. (5.3) and (5.5) would be identical.

The expression for dose at P, including inhomogeneity correction, then becomes

$$D'(d',\phi,g,x) = D(d',\phi,g,x)*s*D_c(d_{max},\phi,g_c)/D_c(d_{max},\phi,0)$$
$$= D_c(d'_c,\phi,g_c)*OCR(d'_c,\phi,g_c,x/H)*s*D_c(d_{max},\phi,g_c)/D_c(d_{max},\phi,0)$$
$$(5.6)$$

where $d'_c = \vec{d'} \cdot \widehat{SP}_c$ and $d' = d_0 + AET*t$.

To compute dose at point P, quantities $g, d_o,$ and t are obtained by calculating intersections of the source-point ray SP with the contours. Projections of distances along the ray on the central axis are obtained by calculating the scalar product of these distances with the unit vector $\widehat{SP}_c$. The AET factor is evaluated by table look-up and interpolation in a table containing AET as a function of d_o, t, g, density and energy. Quantity x, with an appropriate sign, is obtained by forming the vector product of SP with the unit vector $\widehat{SP}_c$.

External inhomogeneities such as boluses are taken into account by modifying the patient's contour.

In the method described above, the following virtual source convention has been employed. It is assumed that the virtual source is at the point of convergence of 50% isodose lines. Two points are chosen on each limb of the 50% curve, and straight lines joining them are drawn and extended towards the source. The point of intersection of these lines and the central axis determines the position of the virtual source. The choice of construction points on the 50% curve is arbitrary. The depth of the first point is 1 cm and depth of the second point in centimeters is numerically equal to $E/4$, where E is energy in MeV.

5.3 MEASUREMENTS

Recent measurements in water were made using a computer-aided dose distribution measurement system. The system, which scans a fan matrix of adjustable divergence and spacing, has been optimized to obtain maximum amount of information with a minimum number of measurements. It consists of a computer driven probe in a water phantom. A 40 cm wide and 3 cm high, thin (6 mil.) Mylar slit window on one side of the water phantom allows for the measurement of central-axis and off-axis depth dose for up to almost zero depths. The system can be connected to the computer (PDP 11T55) directly or over a telephone line. The computer moves the probe to a desired position and triggers the integration of charge in a thin window parallel plate ionization chamber and a reference source in a two-channel electrometer. When

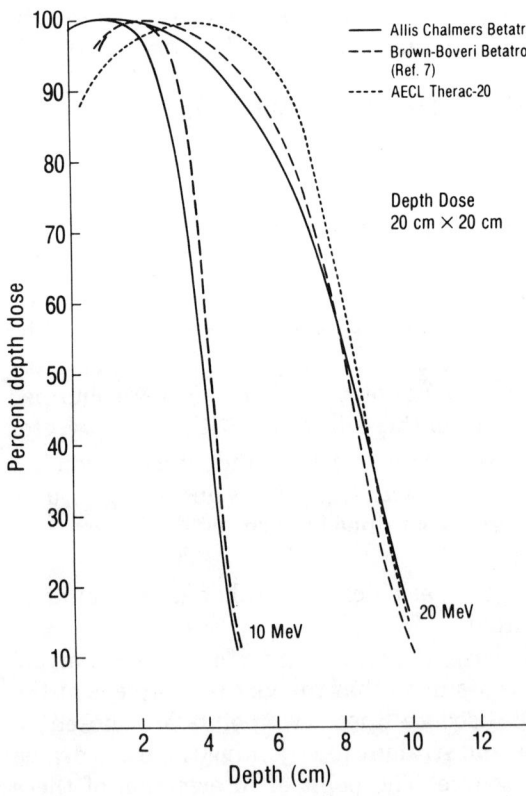

FIG. 5.3. Percent depth dose measurements on the central axis of 20×20 cm electron beams. From Brown–Boveri,[7] Allis Chalmers, and AECL Therac-20[8] accelerators.

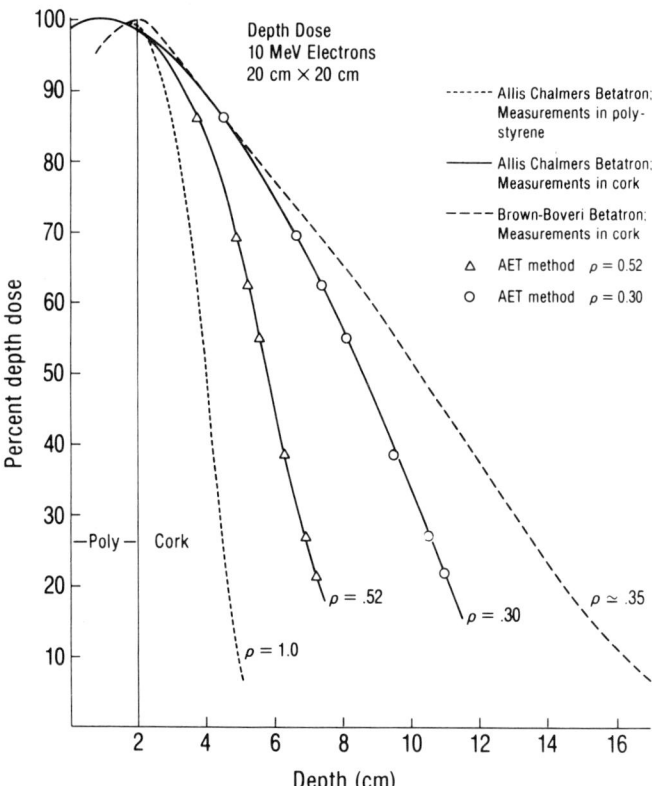

FIG. 5.4. Percent depth dose measurements in lung equivalent cork media of two densities at 2 cm depth compared with AET-calculated depth dose (O,△) and measurements made in cork at 2.5 cm on a Brown–Boveri betatron (Refs. 7,9).

the charge in the reference channel reaches a pre-defined level, it is held until the computer samples the charge accumulated by the exploring chamber and initiates the next motion and measurement cycle. The measured data are analyzed to obtain off-center ratios, percent depth-dose data and isodose charts.

The data measured included central-axis depth-ionization data in water and OCRs for air gaps of 0, 3, 6, and 9 cm for each energy and each field size (defined at the end of the collimator). Central axis ionization in air, with a "build-up" cap, was also measured. For conversion of ionization measurements to dose, the data are multiplied by the ratio of stopping power in water and in air, including correction for the polarization effect.

Cavity ionization as a function of depth was measured along the central axis in polystyrene, cork ($\rho=0.52$ and $\rho=0.30$) and mixed polystyrene–cork phantoms. The probe was a cylindrical parallel plate ionization chamber, with a nominal volume of 0.25 cc, 2.5 mm electrode

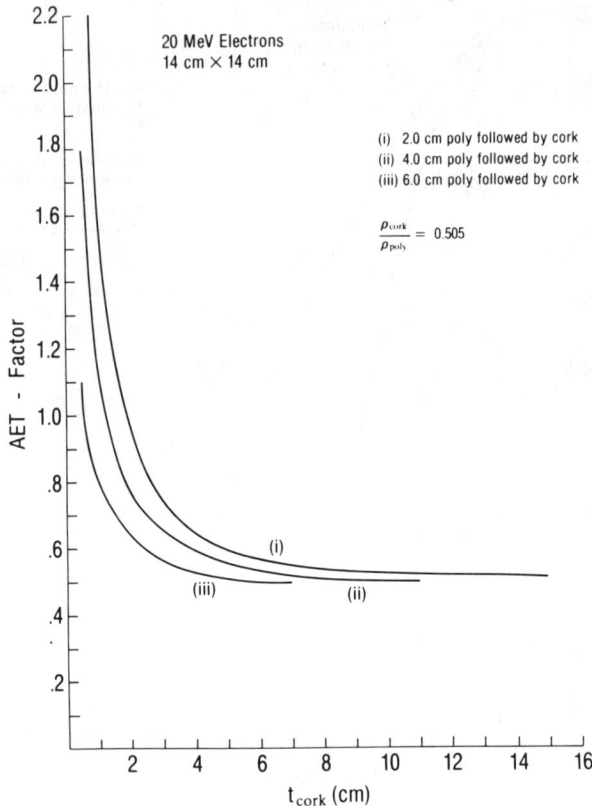

FIG. 5.5. Variation of AET factors with increasing depth into a cork phantom ($\rho = 0.52$ g/cc) for 20 MeV electrons, 14×14 cm: (i) 2 cm interface depth, (ii) 4 cm interface depth, (iii) 6 cm interface depth.

spacing, and 6 mg/cm² front window thickness. Special phantoms were constructed from each material containing cavities which matched exactly the dimensions of the probe. Measurements could be made through cork and across interfaces since all phantom sections were interchangeable. The measurements were made by adding 25×25 cm sheets of arbitrary thickness to the top of the phantom. The SSD was always kept constant, and the machine monitor was used as a dose reference.

Electron beam depth-dose characteristics at the same initial energy as determined from practical range measurements will differ in surface dose, depth of maximum dose, and the rate of fall-off in dose beyond the 90% depth-dose point. These differences are demonstrated in Fig. 5.3 for 10 and 20 MeV electrons from AC, BBC, and AECL accelerators, and are caused by differences in design or absence of field flattening filters and by variations in design of the collimating system. An important question to answer is whether these differences significantly affect heterogeneity correction methodology and data such as AET factors.[3]

Representative ion chamber measurements in cork phantoms of varying density are shown for 10 MeV electrons in Fig. 5.4. The isodoses are sensitive to the decrease in density from 0.52 to 0.30 g/cc as expected and previously reported. With correction for heterogeneities expressed in the form of AET factors, calculations from the standard isodoses in water give excellent agreement with measurements. Careful reporting of experimental technique and materials will continue to be very important to avoid misunderstanding and unresolved discrepancies.

It is clear that corrections for heterogeneities in computation of treatment plans must allow for variation in density and composition of the media, depth into the heterogeneity, field size, and energy. We also observed, as shown in Fig. 5.5, that it is necessary to incorporate quantitatively corrections for the composite nature of the heterogeneous tissues. AET factors are presented for different thicknesses of polystyrene overlying the cork phantom showing less dependence of AET factors on

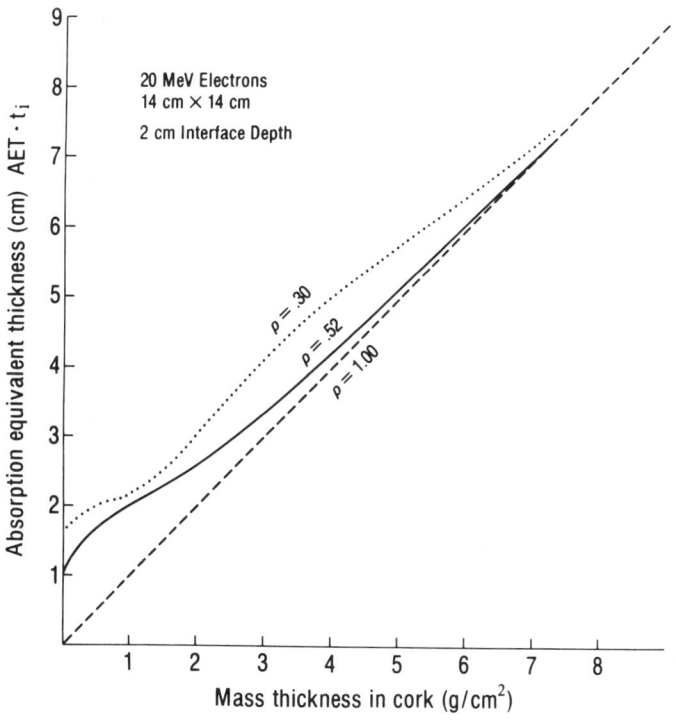

FIG. 5.6. The absorption equivalent thickness, (AET)·(t), for increasing mass thicknesses of cork at $\rho=0.30$, $\rho=0.52$, and $\rho=1.0$ g/cc.

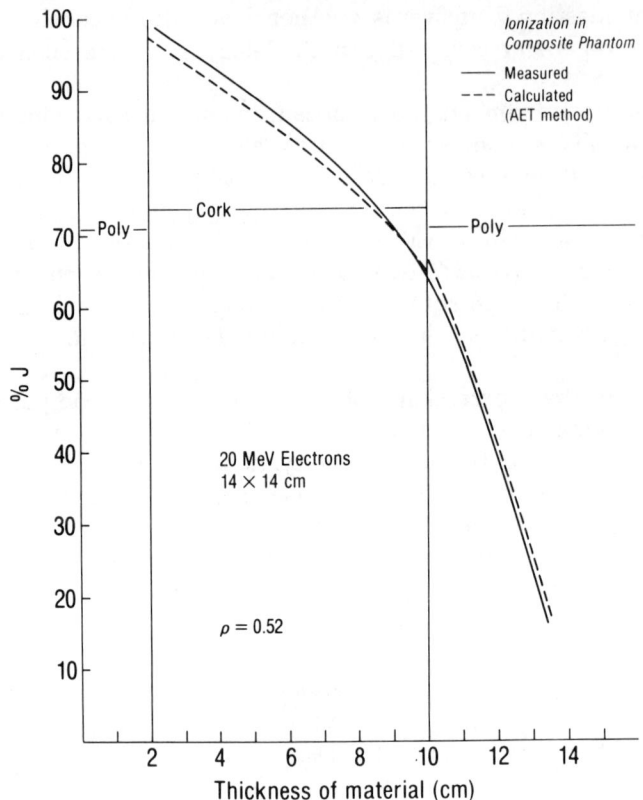

FIG. 5.7. Relative ionization in a composite polystyrene-cork phantom: (—) measurements, and (- - -) AET calculations.

depth at the larger interface depths and in the limit of large cork depths approaching the physical density.

Additionally, as shown in Fig. 5.6, absorption equivalent thickness, $(AET)^*(t)$, where t is the depth into cork, is dependent on the density of the heterogeneity for equal mass thicknesses of the cork. This effect is less pronounced for greater interface depth and is more pronounced for lower energies. The AET method, properly used, and with individual machine data, provides accurate isodose calculations in heterogenous composite media as illustrated in Fig. 5.7.

The method is quite general and is also applicable to other charged particles and neutron beams, as well as to photons.

5.4. REFERENCES

[1] J.S. Laughlin, J. Ovadia, J.W. Beattie, W.J. Henderson, R.A. Harvey, and L.L. Haas, "Some Physical Aspects of Electron Beam Therapy," Radiology **60**, 165–184 (1953).
[2] J.S. Laughlin, "High Energy Electron Treatment Planning for Inhomogeneities," Brit. J. Radiol. **38**, 143 (1965).

[3] A. Dahler, A.S. Baker, and J.S. Laughlin, "Comprehensive Electron Beam Treatment Planning," in *High Energy in Radiation Therapy Dosimetry*, Ann. N. Y. Acad. Sci. **161**, 198–213 (1969).

[4] J.S. Laughlin, *Electron Beams in Radiation Dosimetry*, edited by Attix, Roesch, and Tochilin (Academic, New York, 1968), 2nd ed. Vol. 3.

[5] J.S. Laughlin and J.W. Beattie, "Ranges of High Energy Electrons in Water," Phys. Rev. **83**, 692–693 (1951).

[6] J. Zsula, A. Liuzzi, and J.S. Laughlin, "Oxidation of Ferrous Sulphate by High Energy Electrons and the Influence of the Polorization Effect" Radiation Research **6**, 661–665 (1957).

[7] F. Bagne, "Electron Beam Treatment Planning System," Med. Phys. **3**, 31–38 (1976).

[8] Atomic Energy of Canada Ltd., Private communication.

[9] F. Bagne, private communication.

6. Low-energy electrons

Farideh Bagne, Ph.D.* and Marchant E. Tulloh, M.D.[†]

Dartmouth-Hitchcock Medical Center, Hanover, New Hampshire

6.1 INTRODUCTION

Electron beams in the energy range 1 to 10 MeV constitute a valuable means for treating relatively superficial lesions of less than 3 cm thickness.[1-3] The principal advantage of lower-energy electrons over kilovoltage x rays lies in the particular depth-dose pattern of electrons which exhibits a relatively flat initial plateau in the first few centimeters followed by a rapid fall-off of dose. This is contrary to the exponential absorption of superficial and orthovoltage x rays (Fig. 6.1).

Clinical applications of lower-energy electrons in radiation therapy include irradiation of:
—skin tumors,
—lesions near the body surface,
—the total body,
—the chest wall region.

The following describes briefly the dosimetry of low-energy electrons, with emphasis on lower than 5 MeV energy electrons, and discusses special problems associated with the treatment planning of electrons in the range 1–10 MeV.

6.2. DOSIMETRIC CONSIDERATIONS

The routine dosimetry of electrons with initial energy greater than 5 MeV is generally done using a calibrated thimble chamber. Perhaps, the most commonly employed chamber for this purpose is the Farmer-type 0.6 cc chamber with an air-equivalent (carbon or plastic) wall. At energies below 5 MeV, however, Farmer-type chambers are not suitable for several reasons:

—The relatively thick wall causes lower energy electrons to be scattered and absorbed by the wall, thus producing a large perturbation in the radiation field.

—Because of the large depth of the air volume, the chamber will measure the average dose across a path unacceptably large compared to the range of electrons.

—A fraction of electrons is absorbed by the central collecting electrode which generates a significant negative charge in addition to that

---

*Presently at Duke University Medical Center, Durham, North Carolina.
†Presently at Ottawa Civic Hospital, Ottawa, Ontario.

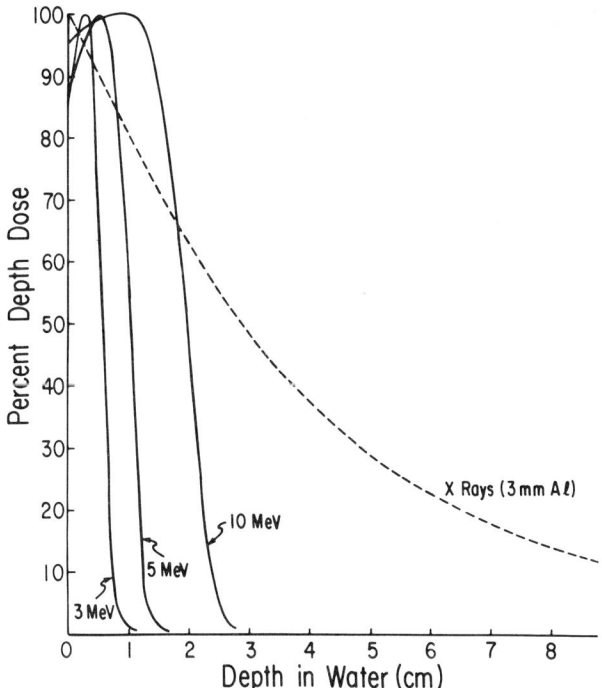

FIG. 6.1. Comparison of the percent depth dose curves for orthovoltage x rays (3 mm Al filter) and 3, 5, and 10 MeV electrons (BBC 45 MeV betatron).

induced by the ionization within the cavity. This problem, however, can be overcome by changing the polarity of potential of the collecting electrode and averaging the chamber readings.

The requirements for an ideal ion chamber for dosimetry of low-energy electrons include the following:

—The chamber must be constructed of tissue-equivalent material.

—The depth of the air cavity in the path of the beam must be small compared to the range of electrons.

—The central electrode must be thin enough so that the polarity effect becomes negligible.

—The connecting cable must be adequately shielded from radiation.

—The lateral dimensions of the air cavity must be sufficiently small to allow dose measurement in a non-uniform field.

—Means must be provided to submerge the chamber in water.

—The walls must be made so as to introduce the least amount of perturbation in the field.

—The geometry must be suitable for in-phantom use.

—The saturation characteristics of the chamber must be adequate for pulsed radiation.

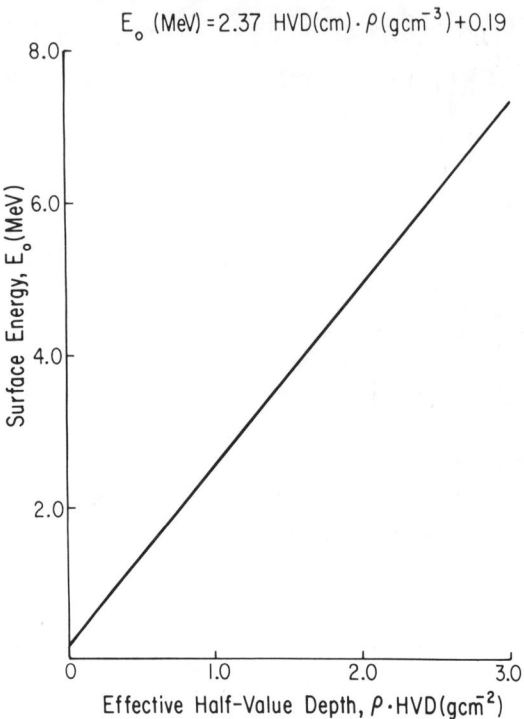

FIG. 6.2. Relationship between the beam surface energy, E_o, and the effective half-value layer in tissue equivalent material.[5]

—The instrument must be rugged, reproducible, yet economical.
—The displacement factor must be either negligible or at least energy independent.

Although several chambers have been designed for this purpose,[4-7] as yet no system exists which statisfies all of the above requirements. The Hospital Physicists Association, in collaboration with the National Physical Laboratory, has designed a low-energy electron chamber suitable for use with solid phantoms.[7] The HPA recommendations on the dosimetry of low-energy electron beams are given in the HPA Report No. 13 "A Practical Guide to Electron Dosimetry Below 5 MeV for Radiotherapy Purposes". The following presents a modified version of these recommendations.

In order to calculate dose to a point in water from the chamber readings, the following equation derived by ICRU [8] is used:

$$D = RNC_{P,T} C_E$$

where D is the dose at the measuring point, N is the chamber energy calibration factor for Co-60 γ rays. $C_{P,T}$ is the temperature and pressure correction factor. The parameter C_E, the overall conversion factor which relates the chamber reading R to dose, is a strong function of beam energy and depth. The beam surface energy, E_o, for low-energy electrons can be estimated from the half-value depth measurements as shown in Fig. 6.2. Figure 6.3 presents the C_E values as a function of energy for a number of depths. As the dose is often measured in a solid phantom rather than water, the tissue equivalent, rather than the actual depths are given.

6.3 TREATMENT OF SKIN TUMORS

Skin tumors are relatively shallow and are conventionally treated with superficial x rays. Skin lesions of 2–3 cm depth, such as certain lesions of the head and neck, can easily be treated with electrons. Treatment planning of these areas may require irregularly shaped fields. The preparation of field shaping blocks has already been discussed in detail by previous speakers. To reduce dose by 95 to 98%, it is sufficient to use a lead shield with a thickness of $\frac{3}{16}$ of an inch.

In order to obtain a uniform dose distribution within the blocked field, the lead shield must be placed such that air gaps between the

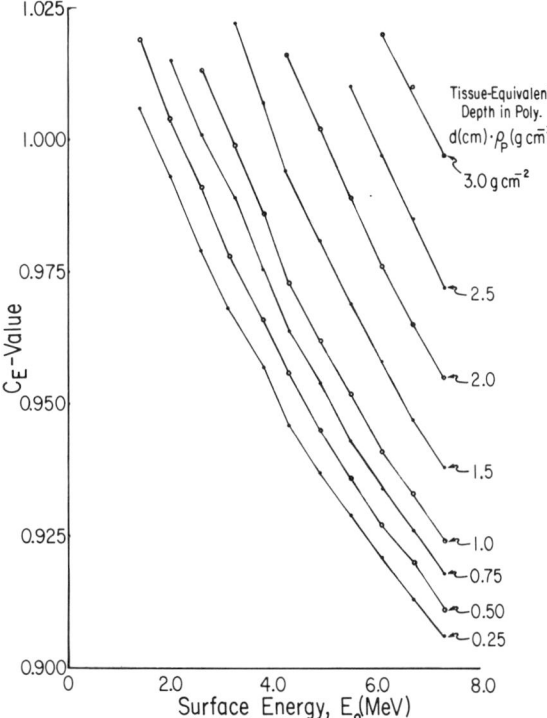

FIG. 6.3. C_E values plotted as a function of surface energy E_o for a number of depths in a tissue-equivalent material. The depths are corrected for the density of phantom ρ.

patient's skin and the cutout are avoided. The uniformity of the blocked field is preferably verified prior to its use for patient treatment. The verification may be accomplished by film dosimetry in a phantom which simulates the anatomical region to be treated.

If the lead shield covers the field by more than 20%, special calibration is required to determine the reference dose for the irregular field. If the shielded region is more than 30% of the total area of the treatment cone, an additional central axis percent depth-dose measurement must be done to confirm the location of maximum dose and to test the validity of the percent depth-dose curve of the unblocked treatment cone for the blocked field.

6.4. TUMORS NEAR SKIN SURFACE

In certain clinical situations, it is advantageous to employ low-energy electrons for the treatment of tumors near skin surface. Examples are lesions of the ear, lip, nose, and cheek. Although the dose decreases rapidly beyond the tumor, there are cases where it is desirable and feasible to protect underlying tissues even further by placing a shield behind the treatment volume. Certain studies have indicated a large increase, as high as 80%, due to electron backscattering from the shield.[9-11] To reduce the backscattering effect, it is possible to replace the commonly used lead shield with a low-Z material such as aluminum[9] or else cover the lead piece with a thin layer of wax.[10]

6.5. TOTAL BODY IRRADIATION

Whole body irradiation for the treatment of certain diseases which cover large areas of the skin can best be achieved by means of low-energy electrons in the range 2–4 MeV. The present discussion, however, will be limited to the available techniques for the treatment of mycosis fungoides.

Mycosis fungoides is a rare form of malignant lymphoma that begins in multiple foci in the skin and gradually progresses to involve various regions of the skin with indurated plaques and tumors.[14] A number of techniques have been used for the treatment of this disease[12-40]; the treatment of choice, however, is considered to be electron therapy.[13] The use of low-energy electrons for the treatment of superficial tumors had been suggested by Trump et al.[35] as early as in 1940.

The technique employed for total body irradiation depends on the type of accelerator, output and energy of the beam, and the design of treatment room. Thus, it is not possible to describe a universally acceptable technique for this purpose. Nevertheless, several precautions are required in treatment planning procedures, as discussed below:

—The skin dose, depth of penetration, and the overall dose distribution, although primarily dependent on the beam surface energy, can be strongly affected by the patient's anatomical cross section, number of fields employed, and the shielding conditions.

—Lead shields of approximately $\frac{1}{16}$ of an inch thick are normally used to protect the eyes from possible cataract formation. The adequacy of eye shields must be verified.

—The percent depth-dose curve must be obtained under treatment conditions and at the correct source-skin distance (SSD), since the placement of absorber in the path of the beam as well as the absence of a treatment cone and the longer source-skin distance, may change the beam penetration considerably. Film dosimetry is generally the preferred technique for these measurements.

—The most important consideration in total body treatment planning is the uniformity of the composite dose distribution. In this case, the simplest and most informative, though not necessarily the most accurate, method for determining dose distribution is film dosimetry as it provides comprehensive data on the total dose received by the irregular surfaces under treatment conditions.[20]

Prepackaged films are cut to the size of the patient's cross-section at various anatomical levels. They are then loaded tightly between slabs of a tissue-equivalent phantom also shaped to the patient's contour. The shaping of the phantom is done by adding tissue-equivalent bolus (see section on bolus). At anatomical levels where the cross section is small, such as the neck and extremities, the corresponding films cannot be readily processed by an automatic film processor. Much information is lost if the films are taped to a large film, as is done routinely in diagnostic radiology and in nuclear medicine. This problem can be solved by leaving a small rectangle on one side of the film in addition to the required area. When the film is placed in the phantom, the protruding section is bent down and taped securely to the side of the phantom so as not to interfere with the surface dose distribution at that location. Subsequent to irradiation the extra rectangular section of the film is taped to a large film, using the special tape made for this purpose, and processed as usual. Little disturbance is produced in dose distribution by the presence of the extra piece, yet the film can be processed by an automatic film processor along with other sections without any extra effort.

To relate the composite isodose distribution to dose at the calibration point, a film is placed in a phantom with the film plane parallel to the beam axis and is exposed to electrons under the calibration conditions (for example, the film is exposed only to a single field). This calibration film, processed along with the other films, is used for dose normalization. The film density at the depth of calibration is assumed to be 100%. All other areas are compared to this point. In this way regions of overdose and underdose are easily delineated. Furthermore, with care, an uncertainty of less than 5% can be achieved.

In order to monitor patient dose, either thermoluminescent dosimeters or film may be used (Fig. 6.4). Based on an extensive study of the various monitoring techniques we have found the following film method to be the most reliable, informative, and simplest for total body patient dose verification.

Thin strips of slow verification type film are cut in the darkroom and inserted in light-tight thin plastic sleeves. Sleeves of various lengths have been made for different anatomical sections. Also cut are strips of

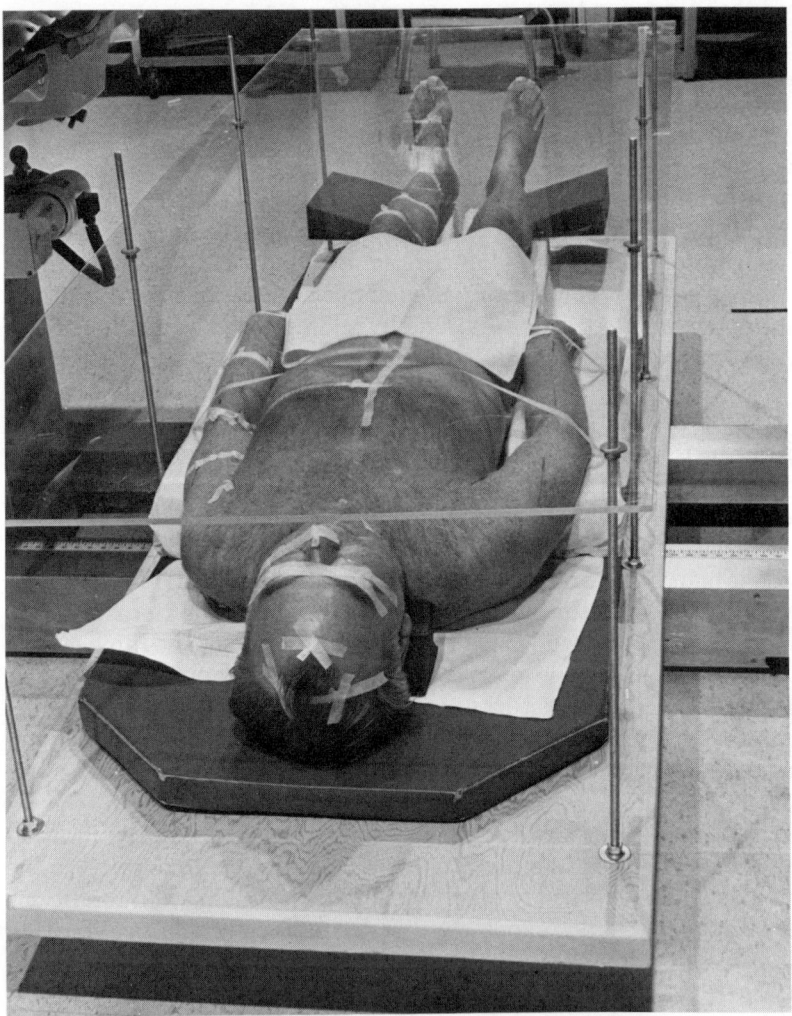

FIG. 6.4. Patient monitoring with TLD. Teflon discs or powder encapsulated in plastic tubes can be used for this purpose.

6. Low-energy electrons

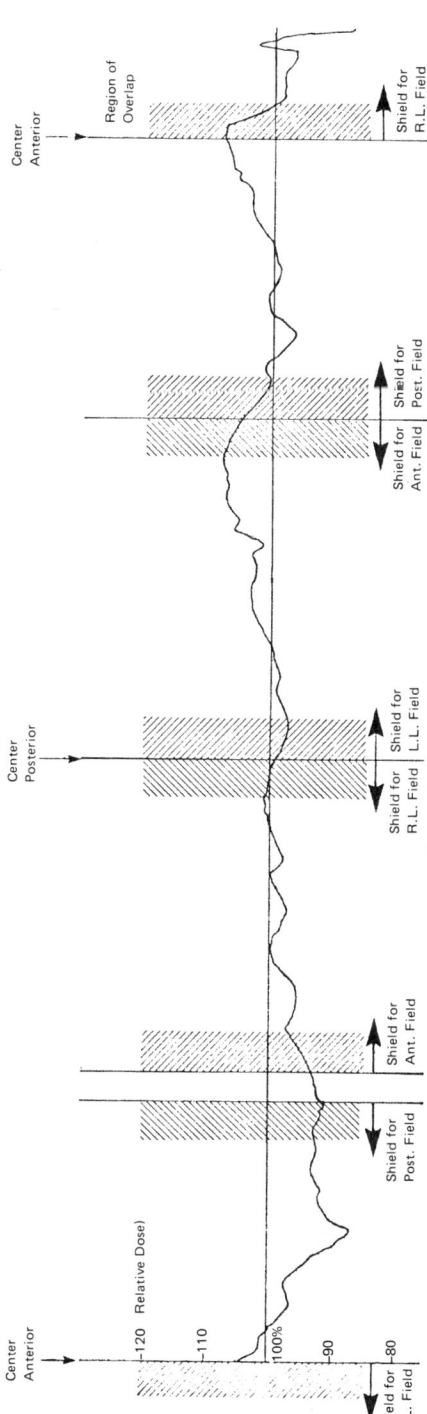

FIG. 6.5. Patient dose verification using film strips. The film was wrapped around the patient's calf. Four-field scanning technique was used. Areas of overlap were shielded with lead. The resulting dose distribution at the depth of maximum dose is shown across the circumference of the patient's calf.

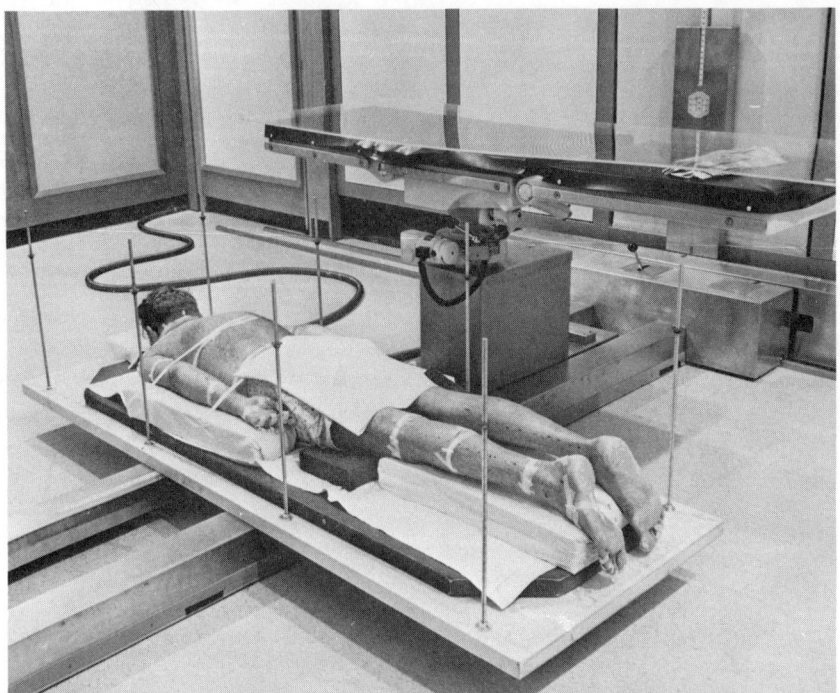

FIG. 6.6. The procedure used at Dartmouth for total body irradiation. Patient lies on a wooden bed placed over the table rail. Sponges are placed under the patient to minimize the effect of the inverse-square law. A four-field scanning technique is used. This figure shows the patient being set up for the posterior field. Subsequent to the setup a plastic tray is fitted over the wooden table which serves two purposes: a) it degrades the beam quality, and b) it provides support for lead shields.

dental wax which are placed over the loaded sleeves. The thickness of the dental wax is chosen so that the film is at the depth of maximum dose. After the patient has been set up for treatment, the appropriate strips are wrapped tightly at the desired anatomical sections which have already been marked. All fields are then treated as usual. We have found an average of 20–25 rads to be the optimum dose for the Kodak Rapid Processing RP/V film. After irradiation, the film strips are processed and a densitometer is used to scan along the strips. An example of the final outcome is shown in Fig. 6.5 for a patient with mycosis fungoides treated with a scanning technique using a modified MeV electron beam (45 MeV Brown–Boveri betatron). The patient was scanned with four fields; the overlapping areas were shielded as shown in Figs. 6.6, 6.7, and 6.8. The nominal source skin distance is 250 cm and the betatron cone is removed. The linearity of the Kodak RP/V film response to dose is shown in Fig. 6.9 for the above beam energy and treatment conditions.

6.6. TREATMENT OF CHEST WALL

The chest wall region is one of the anatomical sites most suitably treated with electrons. Often, however, because of the sharp surface curvature of this region a single field does not provide a satisfactory dose distribution. A compromise is made by employing a series of adjacent angled fields. The resulting dose distribution usually includes localized areas of high and low doses at the superposition of two fields. A common practice for reducing or eliminating such dose nonuniformities is to place tissue-equivalent wedges or properly shaped bolus at the corners of the fields. The following illustrates a technique we have developed for determining the shape of bolus in three-dimensions used for blocking selective portions of the chest wall. The treatment technique is assumed to be a set of angled adjacent fields. Figure 6.10 presents the patient's contour, the area to be treated, and the positions of internal organs. It is desired to cover the entire treatment area with the 80% isodose line.

The necessary steps are as follows:

—The outer contour is traced on two sheets of cardboard and the outlined areas are cut.

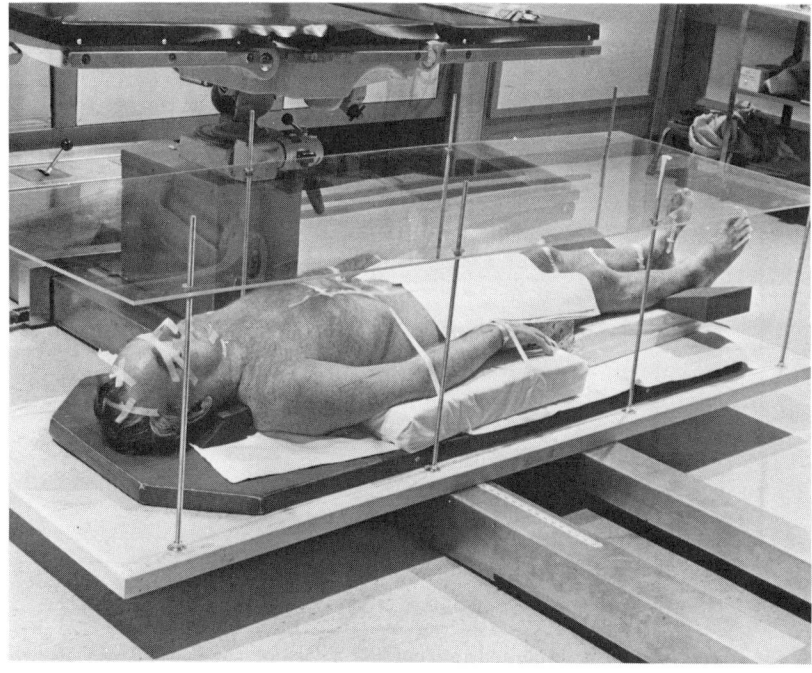

FIG. 6.7. The setup for the anterior field. Lead shields protect the patient's eyes and nails.

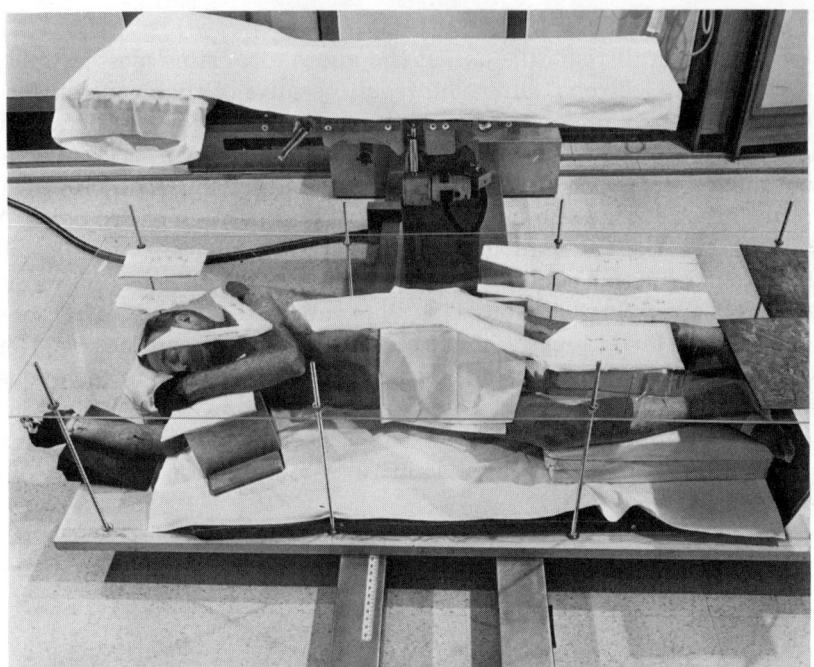

FIG. 6.8. The right lateral field. Lead shields are placed over the plastic tray to protect regions of overlapping dose.

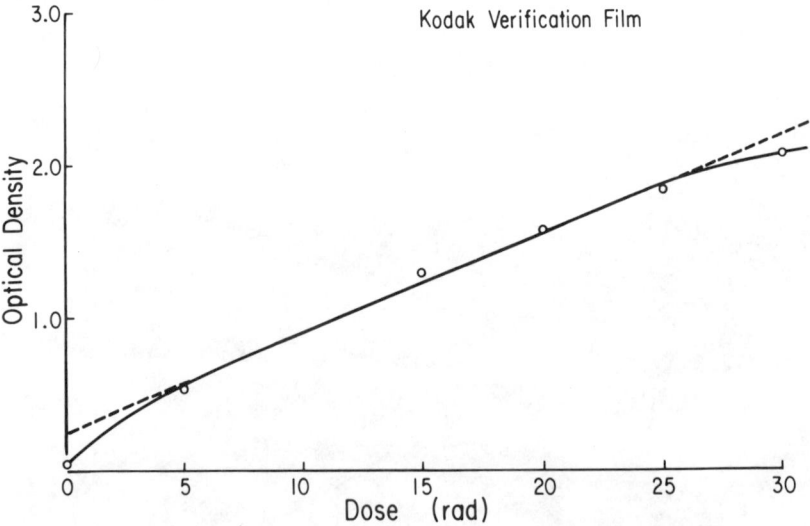

FIG. 6.9. The optical density is plotted as a function of dose for a degraded 5 MeV electron beam. The response appears to be linear in the range 5 to about 25 rads. The film used is Kodak rapid processing verification type.

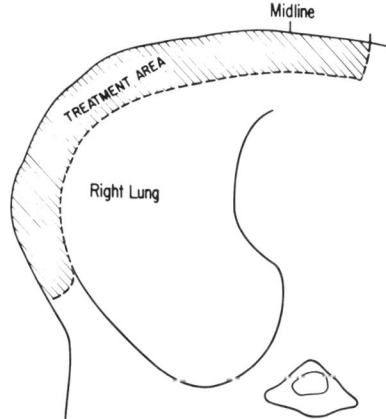

FIG. 6.10. An example of chest wall treatment planning procedure. Patient's contour, the treatment area and the internal organs are outlined.

—The cut cardboard pieces are placed between slabs of a chest-wall phantom and taped securely. We have constructed the chest-wall phantom based on the appropriate anatomical cross section of an average female. The pertinent body structures such as lung, ribs, and tissue are simulated by styrofoam, human ribs, and paraffin. In order to compensate for the differences between the specific patient's anatomy, as portrayed by the cardboard, and the chest wall phantom, we build up the phantom to the appropriate level by using Superstuff (see next section). An example of the chest wall phantom is shown in Fig. 6.11.

FIG. 6.11. Chest wall phantom constructed for treatment planning.

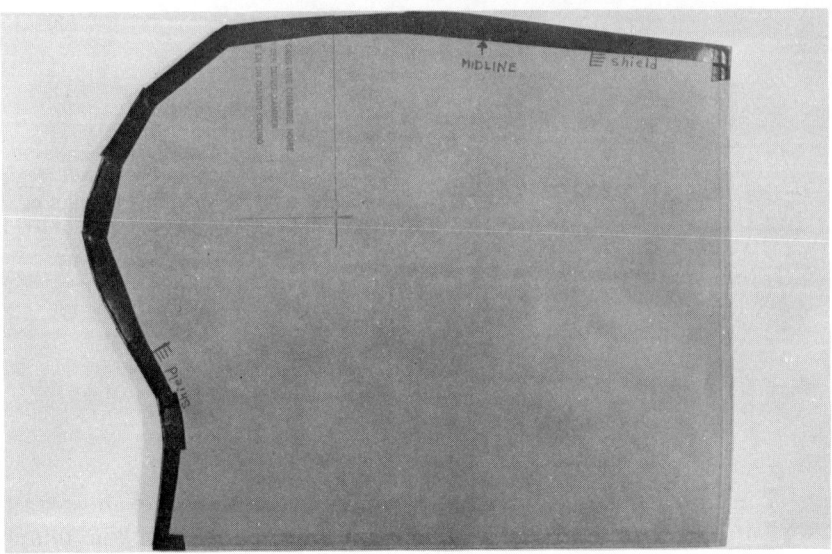

FIG. 6.12. Prepackaged film is cut to the shape of patient's contour. Open edges of the film are taped to eliminate light leakage.

—A prepackaged film such as Kodak rapid processing mammography is cut to the shape of the patient's contour and sandwiched between the two pieces of cardboard (Fig. 6.12). Using a light-tight black tape the open edges of the film are taped to the cardboards to prevent light leakage. The cutting, sandwiching, and taping processes must take place in the darkroom to avoid light exposure.

—The phantom is built up with tissue-equivalent material to match the outline of the cardboards.

—The loaded phantom is set up according to the treatment plan and irradiated. An example of the resulting dose distribution is shown in Fig. 6.13. Overdose areas are evident.

—To reduce dose non-uniformity and to shift the therapeutic isodose line (e.g., 80%) toward the surface so that it extends only to the distal edge of the treatment area, the difference between the therapeutic isodose line and the tumor edge is determined along the ray lines. The derived distances are measured away from the phantom surface, marked on the respective ray lines, and the points are connected.

—The area between the outer contour and the connected points represents the shape of the tissue-equivalent bolus at this cross-sectional level. To obtain a three dimensional bolus the above process is repeated at different anatomical cuts. Cardboard pieces are cut to the shape of the bolus.

—A cast is made of the treatment region and the cardboard pieces are taped securely to its surface at the associated geometrical sections. The bolus is formed by pouring soft wax over the cast and forming it to the shape of the cardboard pieces.

—The adequacy of the prepared bolus is verified by irradiating the loaded chest-wall phantom and bolus under conditions identical to the treatment setup.

The final dose distribution for the above example is shown in Fig. 6.14. The shaped bolus as well as the corner lead shields are shown. Compared to the dose distribution without bolus, Fig. 6.13, this plan offers a satisfactory dose distribution with reduced dose non-uniformity.

In centers where electron beam treatment planning is done by computer, a number of steps in the above procedure are eliminated. In any case, this technique provides a reliable and accurate yet practical means for determining the required bolus shape for adjacent fields and verifying the final treatment plan.

6.7. BOLUS MATERIAL

Treatment planning with electrons often requires the use of bolus to a) negate the effects of surface irregularity, b) increase the surface dose,

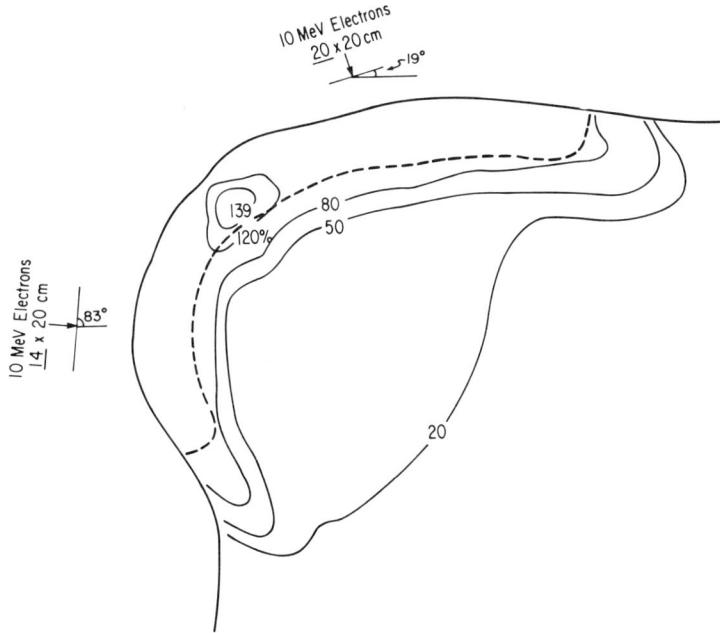

FIG. 6.13 The resulting isodose distribution based on film dosimetry for example given in Fig. 6.11.

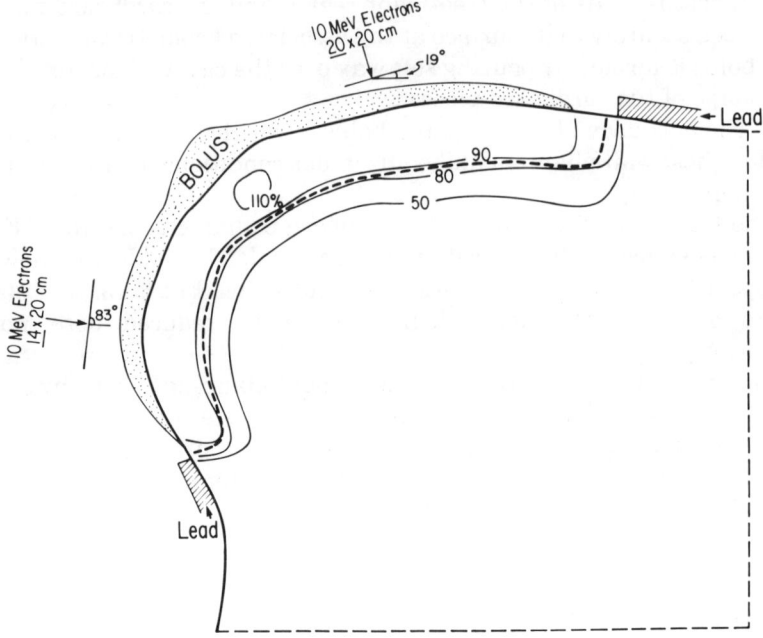

FIG. 6.14. The final dose distribution for the example of Fig. 6.11 using custom made bolus.

and c) reduce the depth of penetration. A number of materials are commercially available for this purpose. The choice of bolus material, however, depends on the particular application. Since most bolus materials are inexpensive, it is advisable to have several different types of bolus available in the department. Table 6.1 presents a list of typical bolus materials used for electron beams, their densities, and the manufacturers.

6.8. REFERENCES

[1] S.M. Jackson, " The clinical application of electron beam therapy with energies up to 10 MeV," Brit. J. Radiol. **43**, 431 (1970).
[2] N. du V. Tapley, *Clinical Applications of the Electron Beam* (Wiley, New York, 1976).
[3] A. Zuppinger, *Frontiers of Radiation Therapy and Oncology.* **2** (S. Karger, Basel, Switzerland, 1968), p. 257.
[4] V. Krishnaswamy, "A thin, flat, parallel plate ionization chamber for electron beam depth-dose measurements," Radiology **109**, 224 (1973).
[5] W.T. Morris and B. Owen, "An ionization chamber for therapy-level dosimetry of electron beams," Phys. Med. Biol. **20**, 718 (1975).
[6] J.W. Boag, "Ionization Chambers," in *Radiation Dosimetry*, edited by F.H. Attix, W.C. Roesch, and E. Tochilin (Academic, New York, 1966).
[7] Hospital Physicists Association, "A Practical Guide to Electron Dosimetry below 5 MeV for Radiotherapy Purposes," HPA Report #13 (1977).

TABLE 6.1. Bolus materials.

Materials	Typical suppliers	Forms	Typical physical densities (g/cm³)
paraffin wax (canning wax)	most supermarkets	block	0.987
Nu-Gel	Smith–Holden, Inc.	powder	1.00
Superstuff	Whammo Corp. 835 E. Elmonte St. San Gabriel, Calif.	powder	1.00
Hygienic pink-base plate wax (type II medium wax)	Smith–Holden, Inc.	sheet	0.99
polystyrene	Cadillac Plastics Boston, Mass.	sheet	1.026
Lucite	Cadillac Plastics Boston, Mass.	sheet	1.25

[8] ICRU, "Radiation Dosimetry: Electrons with initial energies between 1 and 50 MeV," Report No. 21. Int. Com. Rad. Units & Meas. (Washington, D.C., 1972).

[9] J.E. Saunders and V.G. Peters "Back-scattering from metals in superficial therapy with high energy electrons," Brit. J. Radiol. **47**, 467 (1974).

[10] H. Weatherburn, K.T.P. McMillan, B. Stedeford, and K.R. Durrant, "Physical measurements and clinical observations on the backscatter of 10 MeV electrons from lead shielding," Brit. J. Radiol. **48**, 229 (1975).

[11] F. Nusslin, "Electron backscattering from lead in a perspex phantom," Brit. J. Radiol. **48**, 1041 (1975).

[12] W.E. Clendenning, G. Brecher, E.J. Van Scott, and Bethesda "Mycosis Fungoides: Relationship to malignant cutaneous reticulosis and the Sezary syndrome," Arch. Derm. **89**, 785 (1964).

[13] L.Z. Nisce, G.J. D'Angio, and J.H. Kim, "Weekly total-skin electron-beam irradiation for mycosis fungoides," Radiology **109**, 683 (1973).

[14] Z.Y. Fuks, M.A. Bagshaw, and E.M. Farlier, "Prognostic signs and the management of the mycosis fungoides," Cancer **32**, 1385 (1972).

[15] E.J. Van Scott and J.D. Kalmanson, "Complete remissions of mycosis fungoides lymphoma induced by topical nitrogen mustard (HN2)," Cancer **32**, 18 (1972).

[16] G.J. D'Angio, H. Ageto, L.Z. Nisce, J.H. Kim, R. Jolly, D. Buckle, and J.G. Holt, "Preliminary clinical observations after extended Bragg peak helium ion irradiation," Cancer **34**, 6 (1974).

[17] M.I. Smedal, D.O. Johnston, F.A. Salzman, J.G. Trump, and K.A. Wright, "Ten year experience with low voltage electron therapy," Am. J. Roentgen. **88**, 215 (1962).

[18] Z. Fuks and M.A. Bagshaw, "Total-skin electron treatment of mycosis fungoides," Radiology **100**, 145 (1971).

[19] L. Szur, J.A. Silvester, and D.K. Bewley, "Treatment of the whole body surface with electrons," The Lancet, June 30, 1973.

[20] V. Page, A. Gardner, and C.J. Karzmark, "Patient dosimetry in the electron treatment of large superficial lesions," Radiology **94**, 635 (1970).
[21] L. Szur, "The treatment of mycosis fungoides and related conditions with particular emphasis on electron therapy," Brit. J. Cancer **31**, Suppl. II, 368 (1975).
[22] G.J. D'Angio, L.Z. Nisce, and J.H. Kim, "Weekly total skin electron beam therapy for mycosis fungoides and other cutaneous lympomata: further experience," Brit. J. Cancer **31**, Suppl. II, 379 (1975).
[23] J.H. Grollman, S.M. Bierman, R.E. Ottoman, J.E. Morgan, and J. Horns, "Total-skin electron beam therapy of lymphoma cutis and generalized psoriasis: clinical experiences and adverse reactions," Radiology **87**, 908 (1966).
[24] J.A. Levi, C.H. Diggs, and P.H. Wiernik, "Adriamycin therapy in advanced mycosis fungoides," Cancer **39**, 967 (1977).
[25] R. Morrison, "Electron Therapy at 8 MeV," Pro. Ray. Soc. Med., 58.2 (1965).
[26] Z. Fuks, "New Concepts in the management of mycosis fungoides," Brit. J. Radiol. **90**, 355 (1974).
[27] C.J. Karzmark, R. Loevinger, R.E. Steele, and M. Weissbeth, "A technique for large-field, superficial electron therapy," Radiology **74**, 633 (1960).
[28] C.J. Karzmark, "Large-field superficial electron therapy with linear accelerators," Brit. J. Radiol. **37**, 302 (1964).
[29] C.J. Karzmark, "Some aspects of radiation safety for electron accelerators used for both x-ray and electron therapy," Brit. J. Radiol. **40**, 697 (1967).
[30] C.J. Karzmark, "Physical aspects of whole body superficial therapy with electrons," Front. Rad. Ther. & Onc. **2**, 36 (1968).
[31] J.G. Trump, K.A. Wright, W.W. Evans, et al., "High-energy electrons for the treatment of extensive superficial malignant lesions," Am. J. Roentg. **69**, 623 (1953).
[32] M.A. Bagshaw, H.M. Schneidman, E.M. Farber et al., "Electron beam therapy of mycosis fungoides," Calif. Med. **95**, 292 (1961).
[33] G.R. Edelstein, T.E. Clark, and J.G. Holt, "Dosimetry for total body electron beam therapy in the treatment of mycosis fungoides," Radiology **108**, 691, (1973).
[34] T. Kitagawa, "10 MeV betatron electron beam therapy adapted to a case of mycosis fungoides," Am. J. Roent. **88**, 229 (1962).
[35] J.G. Trump, R.J. Van ed Graaff, and R.W. Cloud, "Cathode rays for radiation therapy," Am. J. Roentg. **43**, 728 (1940).
[36] K.A. Wright, R.C. Granke, and J.G. Trump, "Physical aspects of megavolt electron therapy," Radiology **67**, 553 (1956).
[37] H.F. Hare, J.L. Fromer, J.G. Trump, K.A. Wright, and J.H. Anson, "Cathode ray treatment of lymphomas involving skin," AMA Arch. Derm. & Syph. **68**, 635 (1953).
[38] J.L. Fromer, D.O. Johnston, F.A. Salzman, J.G. Trump, and K.A. Wright, "Management of lymphoma cutis with low megavolt electron beam therapy; nine year follow-up in 200 cases," South. Med. J. **54**, 769 (1961).
[39] M.E. Tulloh, Workshop on Electron Beams: Treatment of Mycosis Fungoides, ASTR Annual Meeting, San Francisco, 1975.
[40] P.J. Tetenes and P.N. Goodwin, "Comparative study of superficial whole-body radiotherapeutic techniques using a 4-MeV nonangulated electron beam," Radiology **122**, 219 (1977).

7. Discussion

Discussion Leader, Peter R. Almond, Ph.D.

M.D. Anderson Hospital, Houston, Texas

Dr. Almond

It is apparent from these presentations that electron beam treatment planning physics has many varied and subtle aspects. You have heard almost every speaker this afternoon say that you must measure the parameters of the electron beams on your own machine. I'm not sure that that is where we always want that to be. Perhaps one of the areas we want to look at in the future is some form of optimization, or characterization, or standardization of electron beams, so that information obtained at one center can be used in another center. After all, the electrons produced in Philadelphia are no different from the electrons produced in St. Louis, even though one may come out of a betatron and the other out of a linear accelerator. What matters is what we do with that electron once it gets out of the machine. We use it clinically, and I think we could do more than we have in the past in characterizing these beams. As you know, Brahme and Svensson[1] published in *Medical Physics* last year what, I think, was really the first attempt to try to characterize the beam in some standard notation. I think that we need to look at that approach so that we can get a feeling for why electron beams differ.

Due to the nature of the radiation itself, extra care must be taken in dosimetry, treatment planning and beam stability. I think you realize that for electron beams we are dealing on the whole with much steeper dose gradients within the patient when compared with x-ray beams, and that just small changes in beam energy, body composition, or beam flatness, etc., can have a significant effect upon the dose delivered to the patient. That's really why this subject is so very important. For the x-ray case, we can tolerate larger changes in any of those parameters than we can in electron beam therapy. From the beginning, medical physicists have been aware of these problems, and as you have heard from a number of the speakers, pioneering work in this area has been done by a number of AAPM members. I have made a list of many of these members, realizing that invariably when you make such a list you leave people off. I apologize to those people I have left off—it was purely an oversight, but I would like to mention people like John Laughlin, Lester Skaggs, Larry Lanzl, Bob Loevinger, and C.J. Karzmark. Various speakers have recommended that you go back and read the literature. I'm going to emphasize that again. Go back and read the early literature. Read the paper by Loevinger, Karzmark, and Weissbluth[2] that Jacques Ovadia talked about. It's an excellent article, and it will save you a lot of work if you are doing research on the electron beam and will stop you from rediscovering the wheel. Read the early work by John Laughlin, who covered so much of this area in the early days. Go back and read Lester Skaggs' work, and find out where the $\frac{2}{3}$ and $\frac{3}{4}$ shift for the ion chamber came from.[3] It's all there in the literature, so read it. We owe a debt of gratitude to these people for the sound physical basis upon which present knowledge has been built.

It is clear, however, from the discussion this afternoon, that a tremendous amount of work still needs to be done. As pointed out by Peter Wootton in his opening remarks, the number of high-energy linear accelerator installations has just increased dramatically during the last few years, and it is now imperative that the knowledge we have generated be made available to the increasing number of medical physicists and radiotherapists involved in electron beam radiotherapy. It's not enough to allow these people to use their intuition and knowledge gained with x rays and to transfer that over to electron beams. There are such great differences that at times this can be dangerous. Every year residents come to me, for example, and say: "Why is it with electron beams we have more skin sparing with the low-energy electrons than with the high-energy electrons?" They are thinking, of course, of x rays, where the reverse it true. That can be multiplied into many areas of electron beam therapy.

There are several problems I see, and these are not all of them but some of the things that we are going to have to look at.

For example, moving beam electron therapy, of which you have heard a little today, but that is restricted at the present time to the rather large betatrons produced in Europe. The new generation of linear accelerators are now to come with this capability. I think Toshiba has it, but some of the others are coming out with the capability of doing arc therapy or some form of moving beam therapy with machines that cost considerably less than the large high-energy betatrons, and so it's going to be available, therapists are going to want to use it, and we're going to have to do a lot more in terms of treatment planning with this type of beam.

I think that improvement needs to be made with electron beam collimation and beam control, and I just want to echo again what Jacques Ovadia said of this area and to emphasize again, I just don't think manufacturers have thought enough about electron beam collimation. That is possibly because they have not asked us, and when they *have* asked us they haven't listened. It's partly also that we haven't told them, or if we have told them we haven't insisted upon them making changes. But as you know, you need a machine that is clinically useful. When you get collimators that are too big, or too bulky, or too heavy, or get in the way, there is no need for that. In the literature there are studies reported and work is being done on the basic physics and engineering necessary to produce clinically useful electron beam collimation and beam control.

Just when we think we understand the basic dosimetry, we find new factors which must be taken into account. AAPM has a Task Force working on this and in the next couple of years changes in electron beam dosimetry, not large changes, might come out.

Dr. Bagne talked about low-energy electrons, and they present their own special problems. Incidentally, I see Ken Wright here. The group in Boston have worked with low-energy electrons for years and years and again, some of this can be found in the literature. But what about high-energy electrons? The Europeans are strong on these, but I think Jacques (Ovadia) is the only one that uses beams as high as 33 MeV in the U.S. These beams are known to present special problems to us. The work of Schumacher has been quoted here, and that the European results with very high-energy electrons look amazing is something we should be looking at. In this country most people are content with up to 20 MeV electrons.

In all these areas of research our ultimate aim must be to produce simple, workable, clinical systems, and I want to emphasize simple, workable, clinical. If you go back to the literature for x rays, or indeed electrons, it's full of very complex, although elegant, techniques for doing x-ray treatments. You can't find these being used anywhere around today because they were too complex and difficult. For any system that is going to be used clinically, it must be relatively simple and relatively free from the capability for making errors, and the more complex you have it the more likely you are to make errors. As we work on these systems, I think we must keep in mind that we must make them as simple as we possibly can, consistent with the ultimate aim of delivering the prescribed dose to the patient with the desired precision and safety. We must be careful that we don't get carried away with the complexity of the system that we recommend. There are a lot of areas that we could consider: cost considerations, and also time considerations in applying some of the techniques, should be taken into account. Small clinics with a busy load can't spend time applying a lot of these and, again, we must make sure that what we recommend can be adapted and used safely by these groups. I hope you will get into this discussion.

Does anyone have any questions for any of the speakers?

Question: *Dr. Glasgow*

I have a question for Dr. Bagne. In Fig. 6.13 of your chest wall isodose distribution, would you comment on the amount of beam overlap in that diagram, and then in Fig. 6.14 where you added bolus, would you explain whether or not you set the patient up using the same beam overlap. Did you overlap the beams on the patient surface or did you overlap them on the bolus after it had been added to the patient?

Answer: *Dr. Bagne*

I'm not sure if I understand your question completely, but actually there wasn't any overlap at the surface. That is, the 14 cm field and the 20 cm field were separated.

Question: *Dr. Glasgow*

Then the question is, when you add the bolus do the beam edges meet at the bolus surface or at the patient surface.

Answer: *Dr. Bagne*

At the *patient* surface. We set up the field and then add the bolus.

Comment: *Dr. Ting*

I presented a paper at the 62nd RSNA[4] in which I compared various machines and various tissue density correction factors. The general conclusion from that study was that one does not have to go to very elaborate methods to get first order approximations. One does not have to write comprehensive programs to do all the corrections, which might consist of 20 parameters. If one does 2 measurements of each parameter, one would require 2^{20} measurements. At any rate, my comment was that, if you know the densities of inhomogeneity, one could correct the effect by the ratio of densities, which would lead to reasonable agreement with measured data.

Comment: *Dr. Ken Wright*

A couple of quick comments. I want to thank the Task Group and Chairman for presenting a very wonderful Symposium. The second comment I have is one of concern in using large accelerators with high potential beam current outputs, because of the capability of very high doses in very short periods of time. We have been dealing with large electron machines which have a high energy storage capability. I have always been concerned by this problem, and if you look at the electrons/cm^2 incident or the patient, a total charge of about 10^{-9} Coulombs/cm^2 will deliver a dose of several hundred rads. I would point out that with machines with capabilities of average currents of several hundred microamps, the potential for overdose is very great in short periods of time. I would simply like to emphasize this aspect of safety in using accelerators. Betatrons do not have nearly as great a problem as the large linear accelerators. I think also the body burdens produced in mycosis fungoides treatments have to be very carefully considered, particularly when various methods of treatment are being developed. The depth of penetration times the surface area of the patient which is of the order of 10 000–20 000 cm^2 on the average patient, plus the body burden of the x-ray background, should be considered in the long range point of view in this type of treatment. Many years ago, a Dutch friend of ours (Braams[5]) made some calculations and came up with some figures of what he considered to be tolerable levels which we used to analyze the patient treatments we had done in the previous year, since our early patients had some blood changes produced in them due to the x-ray background. We reduced this background and have had no problems since then.

Comment: *Dr. Ovadia*

My impression was that electron accelerators used for patients have the current limited to a dose that can be set much lower than what could be available for research—is this correct?

Answer: *Dr. Almond*

Most accelerators have dual x ray and electron capabilities. Because you have to run at much higher current with the x rays, there must be some limiting device, either mechanical or electrical, to reduce the beam current when you run with electrons. It is a very good point. There should be some safety device that, if the dose rate exceeds, say, 1000 rads a minute, it should turn the machine off, or at some level you wish to set it at, because you can go up to megarads a second if you are not careful.

Question: *Dr. Suntharalingham*

In terms of electron beam distributions *outside* the central beam, how accurate are these? When Larry (Simpson) showed a measurement technique for off-axis measurements, these were done using ion chambers. Are you using appropriate C_E factors to correct for the actual dose, or are you just taking the ratio of ion chamber readings? And if you are using appropriate C_E factors, do you know what the energy spectrum is at the off-axis points? What sort of errors are we talking about here? I think those of us who are anxious to improve dosimetric techniques, have some questions which remain unanswered, such as when we use ion chambers and film, what type of corrections are necessary?

Answer: *Dr. Simpson*

I didn't make that clear but you're very right. The corrections are those that we're making to the ion chamber measurements that are used for the OCR scans. These are corrected to dose, and we apply the appropriate correction factor for the depth at which that scan is made, and then apply the *same* correction factor all the way across the scan. We feel that using the ion chamber and correcting to the proper depth, we can then use the same correction as we go across the scan—that's an assumption. We don't know any better way to correct the ion chamber reading as we go off axis. We believe that with the tissue-equivalent ionization chamber with a small volume, applying the initial correction to depth and then not changing it for the off axis scans, we're probably then making less of an error, or less assumptions, than we would be if we were trying to use film and correct isodensity, or to use a diode.

Comment: *Dr. Almond*

Dr. Suntharalingham pointed out an area where up until now, we've concentrated to a large extent on the central part of a beam. That is really where most of the work has been done. There has been very little work looking at detector response near the edge of the beam where there is a greater contribution from scattered electrons, and I think some of the discrepancy we seem to observe between diodes and ion chambers which occur in that region is due to the energy response of the detector showing up. I am not aware of any systematic study of this, except that people show this discrepancy from time to time, but I am not aware of anyone who has looked at this closely.

7. Discussion

Question:

I have a couple of comments. First, on the use of CT for measurement of chest-wall thickness. I think that a number of people need to be cautioned about this. We have made measurements on this ourselves with the idea of using this, and have found that on our early scans where we accepted the diagnostic quality films from the Diagnostic Radiology Department, the chest-wall thickness was in error by as much as 100%. This is due to the use of a slow scanner, which thus included chest-wall motion. Also, you can have displacement of the external contour by as much as several millimeters, 3–4 mm, simply due to the way the scan is displayed. So, if you are going to use chest-wall thickness measurements from CT, I suggest you do some phantom studies to determine just how accurately you can measure chest-wall thickness. I would like to ask Farideh Bagne about the way she was doing her film dosimetry measurements. I know she was using tape to seal the film. We have tried it also, but have had some difficulty with the fact that you have an air gap left in there, and that is going to cause some distortion of your isodose curves.

Answer: *Dr. Bagne*

You are absolutely right. Actually, there are two problems. One is that, if you use any type of tape, other than electric tape, when you come to tear off the tape you generate static electricity which will expose the film, creating black spots on the film. You have to be very careful on the choice of the type of tape you use, and we have found that electrical tape is very good for this reason. Secondly, we try to sandwich the film as tightly as we can, and because, as I showed before, styrofoam gives a lot, so we don't have to worry about that as we can make it very tight without any air gaps. You are absolutely right, and we are very aware of that problem.

Question:

I would like to get a consensus or at least some comments from people regarding using CT scans for planning. What form would they like to get the information in? What was shown was a blow-up of the polaroid film. Are there some thoughts on that—I'd like to hear them?

Answer: *Dr. Almond*

I can add the same comment. It's always seemed strange to me that we take a computer generated image, and make an image of that and then image that to put back into a computer. There must be an easier way to do it.

Comment: *Dr. Sternick*

I agree with the statements made that if you have a computer on one end and a computer on the other, then why bother about a life-size image in between. I think that you don't need to do that and it's possible for one computer to talk to the other computer and never generate that life-size picture in between at all.

Comment:

Just one problem with regard to sending from one computer to another has to do with the fact that the information in the CT scan is just too much that you can't handle all that and you need something to reduce the CT scan image to edges, to interfaces, and use *these* as input into the computer for treatment planning. I think that would probably solve the problem a bit better.

Comment: *Dr. Sternick*

There are numerical techniques which one can use to determine edges—one doesn't even have to determine edges, one could use a matrix-CT scan generated, and extract a whole lot of extraneous numbers you never use in the calculations; and that is not a problem.

Question: *Dr. Almond*

Let me ask a question of Dr. Cunningham.

Does it mean that if we have electron beam capability are we also of necessity going to need a computer to do our treatment planning—this is a cost that departments must seriously consider if they want this capability? What cost and what size computers are we talking about?

Answer: *Dr. Cunningham*

Well, I don't think my comment is going to be relevant to your last point, Peter, it's really on the other subject of taking the CT information and using it, rather directly and I think that makes a great deal of sense. I just wanted to say that we have addressed ourselves in a fairly concerted way to that in the recent past. The first thing we do is take the 256×256 matrix that we acquire from the scanner on a magnetic tape and reduce it. In fact we reduce it to a 32×32 matrix. That's a gross degradation of the information, but we plan to go ultimately to a matrix of 64×64 pixel points. Once one has that information in a workable form in the computer, it's really the same problem to determine the position of contours or structures as that of determining the isodose lines. So it's a problem that, in a way, has already been solved or worked on. But, in the determination of the position of structures, for example at the edge of a lung where the density changes from 90% of tissue to 50% of tissue, you find that the structure changes shape quite drastically as you change your criteria on which you define lung.

Dr. Almond

If there are no other questions, I would like to thank again the organizers of this Symposium and all the speakers. Thank you very much.

References

[1] A. Brahm and H. Svensson, "Specification of electron beam quality from the central-axis depth absorbed-dose distribution," Med. Phys. **3**, 95–102 (1976).
[2] R. Loevinger, C.J. Karzmark, and M. Weissbluth, "Radiation therapy with high energy electrons. Part I: physical considerations," Radiology **77**, 906 (1961).
[3] L.S. Skaggs, "Depth dose of electrons from the betatron," Radiology **53**, 868 (1949).
[4] J. Ting and E.S. Sternick, "A comparative study of five tissue inhomogeneity correction algorithms used in electron beam treatment planning," Radiology **121**, 242 (title only) (1976).
[5] R. Braams, "Superficial radiation therapy of large skin areas," Dermatologica **117**, 204–214 (1958).

8. Index

Absorbed dose calibration
   —and collimation, 36,37,98
   —comparison of methods to determine, 15–18
   —determination by
      —chemical methods, 16
      —film dosimetry. 15–19,84–88,103
      —ionization methods, 12,15,16
      —solid state methods, 15–19

Absorption Coefficient Method, 27,54

Absorption Equivalent Thickness (AET), 27,45–47,51,53,54,70–79

Accelerators, linear, 101

Adjacent fields, 20,37–45,99,100

Algorithms, 52–69

Arc therapy with electrons, 60,61,98

Beam
   —characteristics and specifications of, 11
   —energy, 12
   —pencil, 29
   —shaping, 20,27–29

Betatrons, 13,98

Bolus materials, 93–95

Bone attenuation, 27,44,45,50,53,54

Breast
   —field separation for, 37,43–45,99,100
   —irradiation technique for, 34,39,43–45,89–94,99,100

Bremsstrahlung, 12,14,101

Buildup, and surface dose, 12,13,15

Buildup curves, 17

C_E conversion factor, 15,16,75,82,83,102

Calibration, absorbed dose, 16,80–83

Central axis depth dose, 14–16

Chemical dosimetry, 16

Chest wall treatments, 25,34,37,39,43–45,54–56,89–94,99,100,103

Coefficient of Equivalent Thickness (CET), 27,54–56

Collimation, 36,37,98

Computerized tomography (CT), 47,103–105

Computer algorithms
   —analytical, 65–67
      —central axis calculation, 65
      —pencil beams, 65–67
   —empirical, 53–58
      —ACM method, 54
      —AET method, 27,45–47,51,53,54,70–79
      —CET, 27,54–56
      —edge effects, 58–60
      —MAC, Modified Absorption Coefficient, 27, 56–58
   —semi-empirical, 62,64
      —age diffusion, 62,63
      —Czaikowsky, 63,64
      —difference method, 64
      —pencil beam, 63

Cones, 98

Contamination, of electron beams, 12,14,48,49,101

Contours
   —definition of, 22–27,31,46,47,103–105
   —external, 22–25,103
   —internal, 24–27
   —use of CT scans, 22,23,25–27,31,103–105
   —use of transverse axial tomography, 24–26,46,47

8. Index

—use of ultrasound scans, 22,24–26

Curves, isodose, 12,14,36,37,39–51

Density of medium, 27,53,54,75–78

Depth-dose curves, 14–16,33,34

Displacement correction, 82

Dose, absorbed
 —calibrations, 16,80–83
 —central axis depth, 14–16
 —measurement systems, 15,16,80–83
 —perturbations in tissue inhomogeneity, 25–27,44–56,72–78
 —rate, 101
 —surface, 12,13

Dose conversion factor (C_E), 15,16,75,82,83,102

Dosimetry
 —film, 15–19,84–88,103
 —chemical, 16
 —ionization methods, 12,15,16,74–78
 —solid state, 16–19
 —TLDs, 13,16,17

Edge effects, 19,47,58–60

Electron beam
 —basic parameters of, 11,97
 —contaminations of, 12,14,48,49,101
 —energy of, 12,14,33
 —range, 12,33
 —specifications for, 11,97
 —uniformity of, 15,17,18,84

Electron treatment techniques
 —high energy, 99
 —low energy, 80–95,99

Energy
 —determination of, 12
 —loss and multiple scattering, 14,33
 —at surface, 12

Fields, separation of, 20,37–45,99,100

Film dosimetry, 15–19,84–88,103

Inhomogeneous tissues
 —bone, 27,44,45,50,53,54
 —effect on dose distribution, 25,27,44–47,53,54,72–78
 —lung, 26,27,44,45,53–56

Integral dose, 49,50,101

Ionization, and absorbed dose, 15,16,75,82,83,102

Ionization chambers
 —pancake, 74–78
 —transmission, 13

Isodose curves, 19,35,36,74–78,92–94

Lead, for beam modification, 27–29

Linear accelerators, 101

Magnets, scanning, 13,15,29,33

Measurements
 —depth dose, 15,16
 —dose distribution, 74–78
 —isodose curves, 74–78

Modified Absorption Coefficient (MAC), 27,56–58

Multiple fields, 37–45,49

Mycosis fungoides, 84–88,101

Neutrons, contamination by, 48,49

Pendulum therapy with electrons, 60,61,98

Planning of treatment, 27,30,31,33,41

Polarization effects, 70,75

Protective devices, 28,85

Scattering, 13,14,18,27,36,59,60,84

Skin tumors, 83,84

Solid-state dosimetry, 15–18

Surface dose, 11–13

Thermoluminescence dosimetry, 16

Tissue, inhomogeneities, 25,26

Total body irradiation, 84–88

Transmission ionization chambers, 13

Transverse axial tomography, 24–26,46,47

Treatment planning, procedures, 27,30,31,33,41

Ultrasound scans, use of, 22,24–26

Uniformity of electron beam, 15,17,18,84

X rays, contamination by, 14,101

Patty
3388.42

Jenkins & Tubb